The unique contribution made by biological anthropology to human welfare lies in the fundamental understanding it can provide of the dynamic interrelationships between physical and social factors. Topics covered include reproductive ecology and fertility, nutritional status in relation to health, and the effects of pollution on growth. In the later chapters, the concepts of physiological adaptation and Darwinian fitness are explored.

This book will be of interest to researchers and advanced students of biological anthropology, epidemiology, population genetics and physiology, as well as those involved in formulating public policies that affect human health and welfare.

Cambridge Studies in Biological Anthropology 8

Applications of biological anthropology to human affairs

Cambridge Studies in Biological Anthropology

Series Editors

G. W. Lasker
Department of Anatomy and Cell Biology, Wayne State University,
Detroit, Michigan, USA

C. G. N. Mascie-Taylor
Department of Biological Anthropology,
University of Cambridge

D. F. Roberts
Department of Human Genetics,
University of Newcastle upon Tyne

Also in the series

G. W. Lasker   *Surnames and Genetic Structure*

C. G. N. Mascie-Taylor and G. W. Lasker (editors)   *Biological Aspects of Human Migration*

Barry Bogin   *Patterns of Human Growth*

Julius A. Kieser   *Human Adult Odontometrics – the Study of Variation in Adult Tooth Size*

J. E. Lindsay Carter and Barbara Honeyman Heath   *Somatotyping – Development and Applications*

Roy J. Shephard   *Body Composition in Biological Anthropology*

Ashley H. Robins   *Biological Perspectives on Human Pigmentation*

Applications of biological anthropology to human affairs

EDITED BY

C. G. N. MASCIE-TAYLOR

Department of Biological Anthropology,
University of Cambridge, Cambridge, UK

AND

G. W. LASKER

Department of Anatomy and Cell Biology,
Wayne State University, Detroit, Michigan, USA

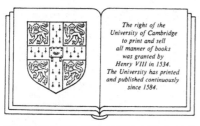

The right of the
University of Cambridge
to print and sell
all manner of books
was granted by
Henry VIII in 1534.
The University has printed
and published continuously
since 1584.

CAMBRIDGE UNIVERSITY PRESS

Cambridge
New York   Port Chester
Melbourne   Sydney

CAMBRIDGE UNIVERSITY PRESS
Cambridge, New York, Melbourne, Madrid, Cape Town, Singapore, São Paulo

Cambridge University Press
The Edinburgh Building, Cambridge CB2 2RU, UK

Published in the United States of America by Cambridge University Press, New York

www.cambridge.org
Information on this title: www.cambridge.org/9780521381123

First published 1991
This digitally printed first paperback version 2005

A catalogue record for this publication is available from the British Library

Library of Congress Cataloguing in Publication data

Applications of biological anthropology to human affairs / edited by
    C. G. Nicholas Mascie-Taylor and Gabriel W. Lasker.
        p.   cm.—(Cambridge studies in biological anthropology: 8)
    Includes bibliographical references and index.
    ISBN 0 521 38112 6 (hardback)
    1. Physical anthropology.   2. Human biology.   I. Mascie-Taylor,
C. G. N.   II. Lasker, Gabriel Ward.   III. Series.
GN60.A67   1991
573—dc20   90-25499   CIP

ISBN-13 978-0-521-38112-3 hardback
ISBN-10 0-521-38112-6 hardback

ISBN-13 978-0-521-01935-4 paperback
ISBN-10 0-521-01935-4 paperback

Contents

Contributors

D. E. Crews
Department of Preventive Medicine and Epidemiology
Loyola University, Chicago, Illinois 60153, USA

P. T. Ellison
Department of Anthropology
Harvard University, Cambridge, Massachusetts 02138, USA

L. P. Greksa
Department of Anthropology
Case Western Reserve University, Cleveland, Ohio 44106, USA

G. D. James
Cornell Medical Center
The New York Hospital, New York 10021, USA

G. W. Lasker
Department of Anatomy and Cell Biology
Wayne State University, Detroit, Michigan 48201, USA

R. M. Malina
Department of Kinesiology and Health Education
University of Texas, Austin, Texas 78712, USA

C. G. N. Mascie-Taylor
Department of Biological Anthropology
University of Cambridge, Cambridge CB2 3DZ, UK

L. M. Schell
Department of Anthropology and Epidemiology
State University of New York, Albany, New York 12222, USA

W. A. Stini
Department of Anthropology
University of Arizona, Tuscon, Arizona 85721, USA

1 Introduction

GABRIEL W. LASKER

There is nothing distinctively new about applying bioanthropology to health and welfare issues. Such applications have taken the form of contributions to medical and dental training and research. In the past, however, the contributions have been conceived of as incidental rather than fundamental: for instance, an anthropologist might be asked about the bodily form of schizophrenic patients because of past literature linking constitutional types to mental diseases and the generally held view that what physical anthropologists do is measure people's bodies. However, a more significant potential contribution of anthropology in the study of schizophrenia may lie in unravelling the interrelationships between ways of life, mental health and bodily development. That is, there are important sociocultural factors in the epidemiology of both schizophrenia and (through diet and activity) of body build.

Human population biology

In this book we propose to emphasize the applications of bioanthropology to broad issues rather than mere applications of anthropological techniques. Because of this, the specifically anthropological aspect of the material may not be clearly distinguishable as such; that is, what makes some of these ideas biological anthropology rather than epidemiology, population genetics or physiology may not seem very substantial. Perhaps it is a matter of emphasis.

Characteristically, bioanthropology is concerned with biological variables in human populations. Thus the concern for the individual – so emphasized in medicine and the other healing arts – is an incident of study rather than an object of investigation for the biological anthropologist. Anthropological study is about the population and for that reason relies heavily on statistical reasoning to define the norms and variations within and among human groups.

Other sciences are heavily statistical too, however, so what is the unique contribution of biological anthropology? In the past anthropologists did field work 'at the ends of the earth' and their work spanned human existence. The conditions of the environment in many of these places during much of this time were unknown. Thus the 'other things being equal' of other sciences could not legitimately be applied. Instead, anthropologists learned to keep an open mind about a wide array of possible influences. They collected and evaluated, in respect to their questions, information that may not have been anticipated to be relevant. Like historians, rather than like chemists, they investigated plausible connections on incomplete evidence and did not confine themselves to the few problems that might have appeared in preset hypotheses.

In studies of biological problems about living human beings, scientific anthropology has by now largely replaced the speculative kind. What has remained from before, however, is a useful breadth of interest which takes into consideration conditions often considered outside their scope by sister sciences concerned with the same issues. Compared with population geneticists, anthropologists may pay more attention to non-genetic influences and be unwilling to lump all the environmental factors into a single category measured by the concept of penetrance. Compared with epidemiologists, anthropologists may be more concerned with varying qualities of life rather than categorical 'affected or not', 'survived or deceased'. Compared with anatomists, anthropologists may be more interested in the causes of morphology – the dynamic processes of evolution and development – even though this dynamic approach envisages transitional forms and obviates clear typologies. Anthropology is holistic, evolutionary, cross-cultural, comparative and population-based.

The chapters have been designed to cover distinct areas, but some issues might almost as well have been broached in a different chapter. The subheadings and the index should help locate particular topics.

In simple terms, human well-being can be reduced to life and health. The threats to life and health are warfare and disease. The bioanthropological approach to these problems is to show that, although they are an aspect of the human condition and hence natural in biological terms, they vary greatly among societies depending on both physical and cultural conditions. It is therefore relevant to reducing rates of mortality and morbidity to know more about the factors in human nature that augment these rates: what is general for the species, what varies genetically among populations, and what is the impact on people of conditions in the physical environment and of circumstances fashioned by the society of the particular social group? A look at the health statistics collected by the World Health Organization and by governments reveals differences in the mortality and morbidity rates between nations. Among human breeding populations defined in social terms, rather than by national boundaries, the variation is even more extreme. Therefore, to understand the variation it is best to look at specific societies and segments of them in the light of their cultures. After allowing for sex and age, the division of societies into sub-areas and social segments may help explain variations in incidence of diseases.

The problems to which biological anthropology is applied

Take a few examples of pressing problems facing the whole human species and consider how biological anthropology looks at them. For example:

1 Extinction of species of plants and animals.
2 Modification of the earth's climate.
3 Mass slaughter in warfare.
4 Overpopulation and exhaustion of natural resources.
5 Accumulation of chemical pollutants.
6 Radioactive contamination from uses, waste and accidents.
7 Genetic degradation of the human species.

The boundaries between the problems are not sharp: thus, overpopulation leads to malnutrition and the risk of war, and warfare increases starvation, the release of chemical pollutants and the risk of radioactive contamination of the whole world. Some of the anthropological aspects of these problems follow.

1 Extinction of species of plants and animals. The most important relationship of human to other species is their use as human food. Are there adequate resources of kinds as well as amounts of food?

Fortunately, human beings are omnivorous and adapt easily to substitution of one food source for another. This capacity is part of the human evolutionary patrimony. Within our memory, anthropologists have demonstrated a much wider range of foods eaten than had previously been known. For instance, until the early years of this century (when the explorer Vilhjamur Stefansson demonstrated – first in the Arctic and then in New York City – that he and his companions could thrive on a meat diet rich in organ foods and fat) nutritionists had believed that human beings needed a vegetarian or mixed diet in order to survive. The higher, non-human primates were then generally thought to be purely vegetarian. Stefansson commissioned a literature search of non-human primate diets and I found, largely from out-of-the-way sources but subsequently repeatedly confirmed in field studies by others, that apes and monkeys also eat insects and even prey on larger game.

Among human groups there are all sorts of diets. Food allergies limit the choice of foods of some individuals, but there is little to suggest that extinction of non-human wild species would have an immediate drastic effect on food availability. The same is not so true of domestic animals and plants. Loss of a species may have grave results, as was evident even from the temporary local loss of the principal crop during the potato blight and famine in Ireland in the 1840s.

Today, one thinks of genetic engineering as the way to deal with such a problem. However, the science of ethnobotany promises access to additional plant species for more general use and tests with new animal species, such as insects, as food have been practically nil.

2   *Modification of the earth's climate.* There is now a prospect of changes in world-wide climates due to the greenhouse effect, and human mobility exposes many more individuals to changed conditions. Can human beings adapt?

Bodily configurations differentially adapt human beings to different climates. Modal body types vary with climate, especially mean temperature in the coldest month in the temperate zones, but also humidity in the tropics. These tendencies follow Allen's and Bergman's 'laws'; that is, that one finds larger body sizes and less surface area per kilogramme body weight in cold climates and the reverse of this in hot ones. However, the situation is complex and there are many exceptions. Geographic variation in human body build is logically related to the need to dissipate heat in hot weather and to conserve heat in cold weather. The logic of nature is that, through repeated elimination by the death of individuals less efficient for these functions, the local gene pools have been culled and have been left, by natural selection, with more of the genes that adapt them to the past climates of the local region. Since past and present climates are correlated – in general, the colder parts of the world having been the colder since the origin of our species – the present geographic distribution of body types would place different kinds of people in the climates where they are best adapted. As a corollary, immigrants would often be less well adapted to the local climate than the sedentes there.

The problem is more complicated, however. For one thing, the principal determinants of body build are not genetic. Height and other linear dimensions of the body are strongly influenced by infant and childhood feeding, and body weight is determined by the ratio of caloric intake to energy expenditure. Different societies have different ways of rationing scarce food; the implications for what is a favourable body build are different if, in times of shortage, food is meted out equally or if larger individuals command larger shares. The relative amounts of lean body mass (the actively metabolizing tissues) and stored fat are also influenced by activity. Thus, overall, relationships of body build to climate and diet may depend on short-term, reversible conditions of food and exercise rather than on genetic tendencies selected by differential survival.

Also, the physiological responses to cold and heat are more significant than body build in the short run. Various studies of physiology and adaptation to temperatures suggest only small differences among human populations; a larger influence seems to be behaviour.

3 Mass slaughter in warfare. What has anthropology to say about war and its causes?

The evidence does not support the proposition that making war is an essential biological trait of the human species. Some oversimplified notions of sociobiology, relying largely on data on selected animal species, pretend that waging war is natural. Of course, it is 'natural' insofar as it occurs. However, a distinguished panel of scientists, including anthropologists Santiago Genoves of Mexico and Philip Tobias of South Africa, concluded in the Statement of Seville that no evidence exists of an inborn human instinct to make war. The killing of others of their own species by some kinds of animals is not closely comparable to human wars and the occasional homicide among hunting and gathering peoples, such as the Inuit, is not warfare. Bioanthropologists and other students of human life need to continue to examine closely each new claim for biological determinism to see whether cultural and other environmental factors are the real cause. Although there is ample evidence against a direct causative link between human biology and war, we have a biology that has allowed us to make war in the past and, in order to avoid repetitions with today's ever more destructive weapons, we need further studies of the ways the proclivity is sometimes channelled in non-destructive directions.

4 Overpopulation. Will we soon be standing several deep on each other's heads?

The study of population growth has largely been the province of demography, a branch of sociology. However, many anthropologists are now engaged in the study. All agree that the population problem can be approached only through the control of fertility, not by increasing mortality. Fecundity, the strictly biological component of fertility, cannot ordinarily be separated from other factors affecting birth rates. Even the concept of 'non-contracepting populations' is theoretical rather than real since all societies have some means of birth control such as the periodic taboos on sexual intercourse, use of abortifacients, coitus interruptus, or by knowledge of modern methods. As with many other issues, the search for culture-free human biology is futile and anthropologists approach the study of such variables as fertility in as wide a variety of societies as possible: that is, in the context of cultural determinants of patterns of marital and extramarital sexual behaviour as well as such biological determinants as sterility and ages at menarche and menopause. Already there is a large body of information from many parts of the world to supplement national statistics on birth rates. Such anthropological data on particular social groups relate fertility to social status. There is little known variation among groups in purely biological (that is, hereditary) fecundity, although the existence of individual genetic differences in sterility, twinning and other components are well established and make some group differences plausible. These are overshadowed, however, by sociocultural factors such as age at marriage and by the interaction of biological with sociocultural factors such as diet and lactation. However, population pressure is often expressed as land shortage and poverty; delaying marriage to acquire bride price or dowry feeds back to reduce fertility. Such mechanisms may be of more influence than the biological mechanisms that reduce fecundity during a period of starvation. Patterns of temporary separation of the sexes may lead to fluctuations in birth rates. For instance, a sharp seasonal variation in numbers of first births occurs in members of one Amish religious sect in which half the marriages occur in the two winter months. More effect on the natural rate of increase in the population is seen, however, where a late age of marriage is traditional, as in Ireland.

One influence on fertility statistics is the unlikelihood of superfecundation. That is, a regular result of pregnancy is not to begin another pregnancy until the previous one is terminated. There is also a biological impediment to another pregnancy during the early post-partum period. Lactation amenorrhea has been well documented by anthropologists. Since the very societies where nursing is extended are often the ones with problems of nutrition and since malnourishment also suppresses ovulation, there is still some controversy about the extent of lactation amenorrhea under various conditions. Thus, those who oppose abortion, sterilization and contraception on allegedly 'moral' grounds may have been led to exaggerate the efficacy of lactation as a way to control population growth. In any case, no one denies the immorality of neglect of unwanted babies, so further evidence on the biological effects of lactation and other aspects of fertility is needed to help guide the formulation of public policy. As a Roman Catholic priest once put it: 'What is not true cannot be right.'

5 *Accumulation of chemical pollutants.* What is the effect on human beings of the increased complexity of our chemical environment?

Efforts at dealing with pollution problems have been aimed largely at reducing contamination by specific chemicals in the workplace. In the United States there are few controls on introducing new chemicals (except into foods and drugs) and in most countries no action is taken until a specific disease is linked to a causative chemical agent, such as berylliosis, the pneumonias caused by inhaling beryllium; there is a high incidence of the disease in exposed individuals, such as those working in the factories that handle it. Tests of a suspected substance by exposing animals to large doses may confirm an adverse effect, but large doses of most chemicals tend to have adverse effects, so such tests may not persuade those making policy decisions to stop exposure of humans. It takes even more effort, cost and time to collect significant evidence of an increased mortality rate when many contaminants are present. On the other hand, smaller doses of one or several substances acting together may show up statistically in the morbidity or growth statistics of entire local populations, and this is where anthropologists can contribute.

However, the joint effects of various chemicals have proven difficult to deal with effectively by social and legal means. Mortality from a number of specific malignancies is considerably elevated in industrial areas where there are multiple contaminants of air and water, such as the Eastern Seaboard Corridor, the Industrial Midwest and the Mississippi and Ohio River Basins of the United States. Although no specific pollutant can be implicated where so many are involved, the problem may be even more important than contaminants in the workplace because it affects whole populations, including the children and pregnant women, of the most densely populated parts of the country. Furthermore, the effects may be especially serious in children and foetuses with rapidly growing tissues and a long life expectancy during which to manifest delayed responses. Specific chemical poisons, such as thalidomide and areas of intense chemical pollution such as the landfill of Love Canal, are well worth investigating, of course, but the more general problem of low-level pollution of large areas by many substances is particularly subject to bioanthropological approaches, approaches necessary for research on relatively small variations in whole populations in the face of numerous compounding variables (age, sex, social class, diet, activities, etc.).

   6   *Radioactive contamination from uses, waste and accidents.* Are there risks of radiation for the future of the species as well as for individuals?
   The relationship of radiation to leukaemia, other cancers and genetic mutation is now understood. Furthermore, there is reason to believe that there is no threshold dose; in other words, no radiation dose, no matter how small, can be considered completely safe. As greater precision has been achieved in studies of the effects, smaller and smaller doses of X-ray and other ionizing radiation have proven to be deleterious. Projection of the dose-damage curve downward shows that even minimum levels of radiation – for instance, the amount coming from radon gas seeping into many modern homes – will lead to some increase in radiation-induced sicknesses. What is the role of biological anthropology in the study of this? The evolutionary perspective states that we are as we are by natural selection and hence are presumably well adapted to conditions as they have been. Increased levels of radiation exposure may be upsetting the balance by increasing the rate of mutation. Most mutations are deleterious to individuals, but a certain low level is advantageous for the survival of the species. An evidence for this is the presence of mutation, but at low levels, in all species that have been studied. Thus, an increase in radioactive exposure with its increase in mutations would not only harm individuals but could also disrupt the equilibrium between individual harm and the good of the species.
   The effects on human biology of other kinds of radiation – light (lasers), sound (airport noise), microwaves (ovens, electronic surveillance) – are so far not well understood.

   7   *Genetic degradation of the human species.* Is our species deteriorating?
   The question of changes in genetic fitness is speculative. The eugenics movement, based on the idea that some genes are good and others bad, has been discredited. The interaction of genetic factors with each other and with environmental factors is complex. What is known about it leads to the conclusion that, for human adaptation, biological similarities are more important than any genetic differences between populations. Through migration, the human gene pool is shared by the whole species,

and heterozygosity within individuals is often beneficial. The interaction with environment is the important aspect of human adaptation, however, and the genetic variation between local populations is apparently less meaningful in this respect. Any study of adaptation of populations must therefore be biosocial and biocultural and cannot be left to a purely genetic analysis. Thus, the subject falls into the anthropological domain. This is not to imply that those trained as geneticists, for instance, cannot make a useful contribution; what it does mean is that they should apply an anthropological perspective, just as anthropologists need to use human genetics when they approach this subject. Baker (1982) emphasized this transdisciplinary nature of the science that equips biological anthropologists to play a central role in studies of the above-listed kinds of issues.

The examples selected

The chapters which follow are roughly in order of the human life cycle: applications of biological anthropology to problems of fertility, childhood development, adult health, degenerative diseases and aging.

Anthropological studies of human fertility began, essentially, as typical demographic studies as far as methods are concerned, but the objects of study were populations of anthropological interest: tribal and other societies with only local or recent vital statistics (Kaplan, 1974). From the outset, such studies were important for questions such as the maximum level of human fertility in populations (Eaton & Mayer, 1953). Furthermore, these very populations exhibit a wider variety of environmental (and probably also genetic) contexts for variations in fertility patterns than do national populations.

It eventually developed that the wide variety of settings for such studies also provided opportunities for more detailed analysis of human reproductive physiology in relation to fertility; that is, fertility as part of the integrated species biology of *Homo sapiens*. It is this fertility and the causes and results of its variability which Ellison reviews in Chapter 2.

In Chapter 3, Mascie-Taylor deals with childhood growth under conditions of a dearth of food. Whereas the poor growth of children often signifies undernutrition and the problems that cause it, the ability to postpone some growth in size helps meet the problem of reduced availability of food for intake. At older ages, the relationship of body size to food required persists, of course, but it is complicated by any relationship that may exist between body size and the ability to acquire food. Estimating lean body mass of whole populations in kilotonnes may be a

10   G. W. Lasker

way to translate body stature, weight, and skinfold thicknessess into
amounts of cereal grain needed (Lasker & Womack, 1975).

Partitioning the variance in young childhood growth into components
for environmental and genetic factors usually shows a large environmental component for height as well as weight, at least in studies where the
environments are diverse. For this reason, human growth has long been a
subject where biological anthropologists have engaged in applied studies,
typically in application of anthropometric methods to the practice of
paediatrics and orthodontics (Tanner, 1955; Garn and Shamir, 1958).
More fundamental applications of anthropological theory are to the
understanding of the evolution of the human growth pattern (Bogin,
1988).

In Chapter 4, Schell raises the question of how growth responds
negatively to chemical or noise pollution. Do such responses occur at
levels of pollution below those resulting in a measurable influence on
mortality or morbidity? In theory, evolution 'views' stunted growth as
preferable to a threat to life and some studies suggest that this actually
happens with several kinds of pollution stress.

The Human Adaptability Project of the International Biological Program (IBP) provided an impetus to studies of adaptation during the
decade 1967–1976. From an anthropological point of view, the most
prominent component was the study of altitude adaptations. That was not
so much because of the millions of individuals whose welfare is affected;
rather, it was because the low oxygen tension encountered by residents at
high altitude is the only stressor that cannot ordinarily be modified by
human management. Hence, altitude is especially useful in studies of
adaptation to stress in general. In addition, the effective research work of
Baker and his students (Baker & Little, 1976), 28 of whom authored a
recent book on the effects of altitude and other stressors (Little & Haas,
1989), has called further attention to the importance of these studies. It is
the example selected by Greksa in Chapter 5 to exemplify adaptation to
stress. The initial impetus for such studies was the IBP; the need for a
review of the subsequent research is now attacked.

As Weiner (1975) pointed out, '[Human] biological adaptation takes
place in an ecological setting which has a predominantly cultural and
technological component. In man Darwinian fitness and physiological
fitness are in many ways sociological phenomena.'

Of course, the two kinds of fitness have very different significance.
Mazess (1975) noted that a great deal of confusion ensues from direct
extrapolation from individual adaptation to notions about adaptation of
populations. In theory, at least, reversible individual adaptations tend to

inhibit hereditary adaptive modifications in breeding populations (Lasker, 1969). Malina discusses the relationship of individual physical fitness to the Darwinian fitness of populations in Chapter 6.

One of the problem areas in which it is difficult to identify the specifically bioanthropological component, and yet potentially one of the most important practical applications of anthropology to human welfare, is in the understanding of the common degenerative diseases of adulthood: hypertension, coronary artery disease, stroke, diabetes mellitus, gout and the malignancies. Such diseases typically have both genetic and environmental factors in their etiology and course. The importance of each causal factor is relative to every other (though not necesarily proportionately), so the factors must be studied in different situations, as anthropologists are wont to do. In fact, the study of similar kinds of individuals in diverse environments occupied through migration is a helpful strategy, as was true in a study of diabetes in Polynesians (Prior & Tasman-Jones, 1981). A variety of such epidemiological studies are reviewed by Crews & James in Chapter 7.

Biological changes are continuous throughout life and, since human beings now tend to live longer than formerly, problems of the elderly have increasing importance in human affairs. We asked Stini, who has been engaged in a longitudinal study of aging of human bones, to address the more general question of aging. He does so in the final chapter (Chapter 8).

The editors are well aware that these are a small selection from a much larger range of applications of biological anthropology. Nevertheless, they are thought to be representative of the kinds of issues in which anthropology has something special to offer. The anthropological approach to these problems is different from that of a professional geneticist, epidemiologist or ecologist, and it illuminates aspects that would otherwise be less well appreciated. One hopes, however, that this very kind of exposition will further disseminate concern for the biology of the whole human life cycle in individuals of varying genetic constitutions and in all sorts of climates, societies and cultures. To the extent that the same and other authors have already succeeded in this, the approach is not confined to anthropology and we hope that it will be even less so in the future.

Summary

Many of the issues of human welfare requiring public action involve small biological effects on large numbers of people. A scientific population approach, rather than the typically individual-centred approach of the caring professions, is therefore needed to mitigate these threats to health. As an example, growth may be a more sensitive indicator of environmental influences than disease. That is, the effects of an untoward influence may show up in an average retardation of growth in a population even when too minor to show statistically significant increases in rates of morbidity and mortality. At least the effects on growth may occur sooner than evidences of disease and thus stimulate earlier efforts to eliminate the source of the problem.

Population policy is a very significant avenue for the application of bioanthropology. Besides its bearing on the issue of war (where 'fewer of them' rather than 'fewer of us' is one of the causes), population pressure has an influence on other aspects of human well-being. That is, long before increase in human numbers will lead to the 'standing room only' that E. A. Ross warned about in the 1920s, it already will have drastically modified the quality of human life. In the following chapters the authors explore some of the issues which exemplify the kinds of applications bioanthropology has to matters of public concern. The findings and interpretations should stimulate the thinking not only of students of such problems, but also of others with a concern about, or responsibility for formulating, public policies that affect health and welfare.

References

Baker, P. T. (1982). Human population biology: a viable transdisciplinary science. *Human Biology*, **54**: 203–20.
Baker, P. T. & Little, M. A. (eds) (1976). *Man in the Andes: a multidisciplinary study of high-altitude Quechua*. Stroudsburg, Pennsylvania: Dowden, Hutchinson & Ross.
Bogin, B. (1988). *Patterns of Human Growth*. Cambridge: Cambridge University Press.
Eaton, J. W. & Mayer, A. J. (1953). The social biology of very high fertility among the Hutterites: the demography of a unique population. *Human Biology*, **25**: 206–64.
Garn, S. M. & Shamir, Z. (1958). *Methods for Research in Human Growth*. Springfield, Illinois: C. C Thomas.
Kaplan, B. A. (ed.) (1974). *Anthropological Studies of Human Fertility*. Detroit: Wayne State University Press. (Reprinted from *Human Biology* (1974), **48**: 1–146.)
Lasker, G. W. (1969). Human biological adaptability. *Science*, **166**: 1480–6.
Lasker, G. W. & Womack, H. (1975). An anatomical view of demographic data: biomass, fat mass and lean body mass of the United States and Mexican human populations. In *Biosocial Interrelations in Population Adaptation*, ed. E. S. Watts, F. E. Johnston & G. W. Lasker pp. 43–53. The Hague: Mouton.

Little, M. A. & Haas, J. D. (eds) (1989). *Human Population Biology: a transdisciplinary science*. New York: Oxford University Press.

Mazess, R. B. (1975). Biological adaptation: aptitudes and acclimatization. In *Biosocial Interrelations in Population Adaptation*, ed. E. S. Watts, F. E. Johnston & G. W. Lasker, pp. 9–18. The Hague: Mouton.

Prior, I. & Tasman-Jones, C. (1981). New Zealand Maori and Pacific Polynesians. In *Western Diseases: their emergence and prevention*, ed. H. C. Trowell & D. P. Burkitt, pp. 227–67. Cambridge, Massachusetts: Harvard University Press.

Tanner, J. M. (1955). *Growth at Adolescence*. Springfield, Illinois: C. C Thomas.

Weiner, J. S. (1975) Opening remarks. In *Biosocial Interrelations in Population Adaptation*, ed. E. S. Watts, F. E. Johnston & G. W. Lasker, pp. 3–6. The Hague: Mouton.

2 Reproductive ecology and human fertility

PETER T. ELLISON

Introduction

Few fields of inquiry draw from as diverse an array of disciplines as the study of human fertility. Contributions come from sociology, public health, medicine, demography, political science, economics, and anthropology, each discipline bringing to bear its own particular perspectives and theoretical agenda. The perspective of the biological anthropologist among these others is unique in two respects: the central position of evolutionary theory and evolutionary history in anthropological thinking, and the commitment to understanding human fertility, its determinants and consequences, as part of an integrated species biology. These two elements give rise to the two primary motivations of biological anthropologists for studying human fertility.

The first motivation is to understand as fully as possible our evolutionary past, both the history of change that we and our phyletic relatives have undergone and the forces and constraints that have shaped its course. Understanding the reproductive biology of our species is fundamental to that effort, since in its essentials natural selection can be broken down into variability in the processes of birth and death. If we can fully comprehend the way in which our fertility is regulated by physiological, ecological and social mechanisms, we will be in a better position to elucidate critical junctures of human evolution, such as the transition to subsistence horticulture, the transition to cooperative hunting and gathering societies, or even the original diversion from other hominoid lines.

The second motivation is in some ways the converse of the first: to understand our species in the present as the product of our formative evolutionary past. That is, as distinct from the often eidetic perspective of medicine or the mechanical perspective of physiology, the biological anthropologist views the apparatus and process of human fertility as shaped by the action of natural selection and searches for the logic of ultimate functionality that can provide a theoretical framework unifying this and other aspects of human biology.

As a result of these two motivations and the basic tenets of the discipline underlying them, the most important contribution of biological anthropology to the study of human fertility is theoretical cohesion, providing a connection to the overarching framework of evolutionary biology.

A quiet revolution has occurred in the last 30 years in our understanding of human fertility which has brought the perspective of biological anthropology to the forefront of the field. The previous view, directly descended from Thomas Malthus, postulated a constant and prolific biology manipulated and constrained by culture, with all the important determinants of variance in human fertility deriving from cultural conventions, societal pressures or individual choice. We now appreciate more fully the degree to which human reproductive physiology is naturally variable and human fertility biologically determined, and the extent to which that determination can be accounted for by evolutionary theory. Rather than the empirical quagmire that characterized the state of the field in the 1950s, we are now well on the way to a theory of natural human fertility. Such an organized theoretical perspective is necessary to support practical applications of knowledge in this area, as, for instance, in anticipating important policy issues in developing societies, or informing clinical perspectives on fertility and infertility. However, as discussed below, scientific knowledge can only support and inform practical decisions concerning policy and intervention. It can never dictate a particular course of action or validate a given policy. The purpose here is to describe the emergence of the new theoretical perspective of human reproductive ecology from its conceptual beginnings to the status of current research, noting some of the implications it holds for issues of public health care and population policy.

The conceptual pioneers

Despite the profusion of empirical studies, the basis of the revolution in human fertility research has been essentially conceptual, with three contributions playing a particularly formative role. Kingsley Davis & Judith Blake (1956) provided an analytical framework for understanding fertility determinants that focused attention for the first time on the biology of reproduction and its constraints. Louis Henry (1961) introduced the definition of 'natural fertility' and the methodology of birth interval analysis, but most importantly provided evidence to suggest the

central role of lactation in regulating natural fertility. Rose Frisch (Frisch
& Revelle, 1970; Frisch & McArthur, 1974) introduced controversial
hypotheses on the relationship of nutritional status and menstrual function, but was also the first explicitly to evoke the agency of natural
selection in designing a reproductive system responsive to key elements
of the environment, and in so doing can be said to have inaugurated the
field of 'human reproductive ecology'. Three decades of research, though
vastly improving our mechanical understanding and quantitative appreciation of the regulation of natural fertility, have not yet crossed the
theoretical frontiers established by these pioneering contributions. Yet
with this conceptual revolution have come heated debates over policy
implications (Rosa, 1976; Frisch, 1978a; Bongaarts, 1980; Population
Reports, 1981; Zeitlin et al., 1981; Bongaarts, 1982; Frisch, 1982; WHO/
NRC, 1983; Bongaarts, 1985; Hodgson 1985a,b; Jelliffe & Jelliffe, 1985;
Konner, 1985; Scott & Johnston, 1985; Wood, 1985), emphasizing that,
in this area as in so many others, basic research provides no haven from
the immediacy of human affairs.

   As sociologists interested in the determination of human fertility by
sociological variables, Davis & Blake were heirs to a legacy of frustration.
Analysis of human fertility with respect to economic status, religious
training, educational level, and endless other supposed determinants
had, over the course of decades of research, yielded only confusing and
contradictory results. Davis & Blake sought a remedy for this conceptual
confusion by directing their attention first to the biological facts of human
reproduction and by trying to identify, exhaustively, those variables but
one step removed through which any more remote sociological determinant would have to operate to have an effect (Davis & Blake, 1956). They
identified 11 such possibilities, grouped in three categories: variables
affecting the formation and dissolution of reproductive unions, variables
affecting exposure to sexual intercourse within reproductive unions, and
variables affecting the probability of conception and pregnancy outcome.
They termed theirs a list of 'intermediate variables', i.e. intermediate in
causation between the sociological determinants of interest, and actual
fertility.

   The heuristic value of this approach was quickly appreciated and
absorbed by anthropologists. In 1962 Moni Nag published his comparative study of 'Factors affecting fertility in non-industrial societies' (Nag,
1962) which sought to apply Davis & Blake's intermediate variables to
the cross-cultural data of the HRAF files. But, though it was originally
intended as an aid to sociologists, Davis & Blake's conceptual framework
ultimately found its closest adherents among demographers, who took up

the task of refining the list of intermediate variables, modelling the theoretical sensitivity of human fertility to each of them, and searching for data to underpin those models. Though the label has changed from 'intermediate variables' (signifying a primary focus of attention on the more remote sociological determinants of fertility) to 'proximate determinants' (signifying a new focus of attention on reproductive biology), recent efforts such as Bongaarts and Potter's (1983) are direct intellectual descendants of Davis & Blake's pioneering contribution.

Yet nowhere in their exhaustive list of intermediate variables, nor in their discussion of their application, did Davis & Blake mention what has now become recognized as one of the principle determinants of natural fertility: lactational subfecundity. It was the French demographer Louis Henry (Henry, 1961) who finally focused scientific attention on lactation with his 1961 paper on 'natural fertility'. Defining 'natural fertility' as fertility in the absence of reproductive behaviour 'bound to the number of children already born', Henry was able to show that populations in this category demonstrated remarkably similar patterns of age-specific fertility, though they differed substantially in level of fertility. Data for such analyses had only recently become available through the coming of age of historical demography and new research in developing countries. Yet for populations as diverse as seventeenth-century Geneva, French Canada, Hutterite communities in America, European families in Tunis, Indian villages, pre-war Japan, and post-war Taiwan, the pattern of female fertility by age relative to the 20- to 24-year-old women was remarkably constant, declining only slightly through age 34, and then declining sharply through the next two 5-year age groups. This pattern contrasted with the more continuous decline from the fertility of the youngest age group demonstrated by controlled fertility populations. Henry was further able to establish that the age-related decline in fertility in natural fertility populations was a product both of increasing numbers of sterile women at older ages and of longer interbirth intervals for those still fertile.

However, although the pattern of age-specific fertility was shared to a high degree among different natural fertility populations, the level of fertility varied over a wide range, the highest fertility populations having nearly three times the total fertility rate of the lowest. This variability, Henry reasoned, must be attributable to one of three causes: variability in the 'idle period' between the conception of one child and the reestablishment of both ovulation and sexual relations; the probability of conception after the idle period is passed; and the frequency of miscarriages and still births. Of these, Henry nominated the first as most in line with the sparse data then available, and suggested that the true cause lay in variable resumption of ovulation associated with differing breastfeeding practices, more than in variable resumption of sexual relations due to postpartum taboos. Yet even this explanation was ultimately deficient, since:

> Ovulation is resumed very rapidly among some women even though they are suckling their babies. With others some time elapses after confinement before it is resumed but still before the infant is weaned; and with still others, it is resumed only after weaning. (Henry, 1961: pp. 90–1)

It is by understanding how women fall into these different categories, Henry felt, that we would finally understand variability in levels of natural fertility, an understanding which would be ultimately 'physiological'.

In 1970 Frisch & Revelle introduced a hypothesis relating the onset of menstrual cycling to female nutritional status. The exact form of the hypothesis has evolved over the succeeding decades from initial reliance on body weight to a more sophisticated and derived notion of body composition as the key independent variable, and to include the maintenance and resumption of adult menstrual function as well as its onset. The historical development of the hypothesis and its contingent criticism is itself revealing (cf. Frisch & Revelle, 1970; Frisch, Revelle & Cook, 1971; Johnston, Malina & Galbraith, 1971; Frisch, Revelle & Cook, 1973; Frisch, 1974, 1976, 1978a,b, 1982; Frisch & McArthur, 1974; Johnston *et al.*, 1975; Billewicz, Fellowes & Hytten, 1976; Cameron, 1976; Trussell, 1978, 1980; Reeves, 1979; Bongaarts, 1980, 1982; Ellison, 1981a,b, 1982; Menken, Trussell & Watkins, 1981), but the fundamental core of the proposition has remained unchanged. As enunciated by Frisch in her most recent writings, the hypothesis states that the onset and subsequent maintenance of regular menstruation depends on first acquiring and then maintaining a minimum level or percentage of body fat, that the minimum for maintenance is somewhat higher than than for adolescent onset, that these minimum body fat levels are necessary but not sufficient requirements for regular menstruation, and that, of course, individual variation exists in the values of these minima for different women.

Controversy has surrounded the Frisch hypotheses from the start, with the most substantive criticism directed at the various methods of data analysis employed and the lack of consideration of alternate hypotheses (see references above). But the most compelling aspect of the hypotheses, and one that remains unscathed by such criticism, is the evolutionary logic on which they are based. Reproduction, Frisch argues, is an energetically expensive undertaking for a female, and natural selection should have moulded female reproductive physiology in such a way that pregnancy would be avoided unless the probability of successful reproduction was sufficiently high. To the extent that the probability of successful reproduction depends on sufficient energy availability, female reproductive physiology should show sensitivity to essentially ecological variables. This is not the only form in which evolutionary logic can be incorporated into an understanding of human fertility, and several alternatives will be mentioned in the course of this review, but since its introduction by Frisch no serious discussion of human fertility and its regulation has been able to ignore the presumptive agency of natural selection in shaping our reproductive biology.

Like explorers, these pioneers established the boundaries of a field of inquiry but left most of the territory uncharted. The detail that has now been painstakingly added to the map, especially in the past 10 years, has enriched our understanding of the mechanisms of human fertility regulation to the point where initial steps at middle level theory formulation are now possible. An important focus of research has been the understanding of human ovarian function, which also provides the point around which theory has begun to coalesce. Controversies remain in many areas, nonetheless, with particular discrepancies between the results of demographic analyses and physiological studies.

Lactational subfecundity

The last 20 years have seen the notion of lactational subfecundity evolve from an old wives' tale to a virtual principle of population science. Little progress was made in the decade after Henry's paper on sorting out women who resumed cycling rapidly post-partum from those who did not. Lactation did not seem to produce a clear answer since, especially among western women, menses resumed in 5 or 6 months on average even among breastfeeding mothers. In the developing world, and in historical data, a clearer effect was discernible (Tietze, 1961; Jain *et al.*, 1970; Van Ginneken, 1974; Masnick, 1979). Even then, a firm relationship to fertility was not always apparent. Using data from the Khana study, Potter was able to demonstrate a significantly longer median waiting time to resumption of menses among Punjabi mothers of surviving offspring than among women whose offspring died within the first month, were stillborn, or aborted prior to term (Potter *et al.*, 1965*a,b,c*).

The difference was presumably attributable to the suppressive effect of lactation on ovarian function. Yet the differential in terms of interbirth interval was substantially reduced (Wyon & Gordon, 1971).

Delvoye and his colleagues provided crucial refinement to the analysis of lactational subfecundity by considering the time-course of prolactin in the blood, the hormone already thought to be responsible for suppressing ovarian function during lactation, in response to a nursing stimulus. A rapid rise in blood levels occurs almost immediately when nursing commences, followed by a somewhat slower decline after nursing ends (Tyson, 1977). Repeated, short nursing bouts, they argued, would cumulate individual prolactin surges and result in a higher average prolactin level through the day than a few, widely spaced bouts. This two-step mechanism, between nursing frequency and prolactin levels, and between prolactin and ovarian activity, was then documented by Delvoye and his colleagues in a study of women in Bukavu, Zaire (Delvoye et al., 1977, 1978). Among these Central African women in an urban setting, a nursing frequency of at least six times a day seemed sufficient to maintain elevated average prolactin levels for over a year, while nursing less than four times a day was associated with a return to non-nursing prolactin levels within 6 months. Resumption of mentrual cycles was found to be closely related to the prolactin decline (Delvoye et al., 1978).

Delvoye's data were cross-sectional and left several unanswered questions. For example, during the second post-partum year, serum prolactin levels fell steadily despite continued nursing and could no longer discriminate menstruating from amenorrhoeic women (Delvoye et al., 1978). Furthermore, the nature of ovarian activity in menstruating mothers seemed to change from the first to the second year, as evidenced by higher average oestradiol and luteinizing hormone levels.

Many of these gaps in the story were soon filled in by the elegant studies of McNeilly, Howie, and their colleagues in Edinburgh (Howie et al., 1982a,b) (cf. also Perez et al., 1971, 1972; and Brown, Harrison & Smith, 1985). Howie and McNeilly followed 27 women longitudinally through the post-partum period, recording data on nursing frequency and pattern, the introduction of supplementary foods, prolactin levels, resumption of ovarian activity as indicated by steroid levels, and resumption of menstruation. They confirmed Delvoye's results in all important respects and added more detail to our understanding. As suspected by Delvoye, the introduction of supplementary foods could be seen to be directly related to a decline in nursing frequency and serum prolactin, with resumption of ovarian activity and menstruation following soon thereafter. Also as suspected by Delvoye, a variable time-lag could be

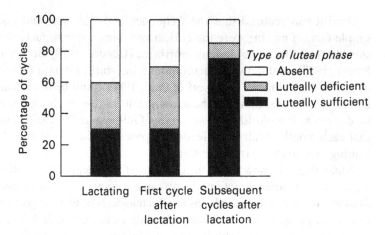

Fig. 2.1. Distribution of types of luteal phases in menstruating Scottish women during and after lactation (McNeilly *et al.*, 1982).

observed between the resumption of ovarian activity and the establishment of regular, ovulatory cycles. A period of diminished fecundity normally persists during lactation even beyond the resumption of menstruation due to anovulatory and luteally deficient cycles (Fig. 2.1). In many respects, the reestablishment of regular cycles was reminiscent of the initial establishment of regular cycles at adolescence, though in a compressed period of time. This analogy w:s strengthened by evidence showing similar changes in the hypothalamic response to oestradiol in premenarcheal adolescents and amenorrhoeic lactating mothers (Baird *et al.*, 1979).

These results still fell far short, however, of accounting for the full range of variability in natural human fertility documented by Henry. It remained for anthropologists to demonstrate that lactational suppression of ovarian function could be extended for years *if* the frequency and intensity of nursing remained high. Konner & Worthman (1980) provided evidence that gonadal steroid levels continued to be suppressed by frequent nursing among !Kung San women of Botswana for at least 2 years. Wood, Johnson & Campbell (1985), in a study of lactation and birth spacing among the Gainj of New Guinea, found that serum prolactin in nursing mothers showed no significant decline with time postpartum independently of the effects of nursing pattern, suggesting that lactational subfecundity could theoretically be extended indefinitely.

Once it was realized that the frequency and intensity, and not the simple fact, of nursing were the critical variables, more robust relationships between lactation and amenorrhoea (Corsini, 1979; Bongaarts & Potter, 1983), and between lactation and interbirth interval (Van Ginneken 1978; Goldman, Westhoff & Paul, 1987) could be demonstrated across populations. In one of the more recent analyses of this kind, based on data from the World Fertility Survey, Goldman *et al.* (1987) estimate that each month of full breastfeeding apparently adds 0.8 months to the waiting time to the next conception.

While there is now wide acceptance of the suppressive effect of lactation on ovarian function, many related issues are still in dispute. Studies disagree, for example, as to which aspects of nursing pattern are important in mediating the effect. Candidates have included frequency (Delvoye *et al.*, 1977), child's age and interbout interval (Konner & Worthman, 1980), 24-hour total duration and frequency (Ojofeitimi, 1982), mean and variance in frequency (Wood *et al.*, 1985), longest night-time interbout interval and 24-hour duration (Gross & Eastman, 1985), and mean and total duration (Vitzthum, 1989). Many of these differences may ultimately be explained by differences in definitions and protocols of data collection. Definitions vary greatly, for example, as to the minimum time off the nipple separating two suckling bouts (Konner & Worthman, 1980; Wood *et al.*, 1985; Vitzthum, 1989).

More importantly, we still have little understanding of the variability in the suppressive effect of lactation, both between individuals and between populations. For example, as will be discussed below, there is good evidence that early- and late-maturing women vary in their levels of ovarian function well into adulthood, yet there are to date no studies of lactational subfecundity with respect to menarcheal age. Nor do we have a sufficient understanding of the relationship between lactational subfecundity and chronological age. Goldman *et al.* (1987) demonstrate a U-shaped relationship between waiting time to next conception and age after controlling for median duration of lactation across 20 countries, but it is unclear whether this derives from an enhanced suppressive effect of lactation at extreme ages or simply lower ovarian function independent of lactation. Furthermore, the necessary restriction of the sample to closed birth intervals leads to an underestimate of the contribution of age to the modulation of fecundity. Nor have the effects of disease burden or workload on lactational subfecundity been adequately studied.

The potential interaction of lactation and nutritional status is currently the area of greatest dispute. Delgado *et al.* (1979) reported a significant negative association for Guatemalan mothers between maternal nutritional status in the third trimester of pregnancy and the length of post-partum amenorrhoea. Lunn *et al.* (1980), in a study of Gambian women, showed that post-partum prolactin profiles varied seasonally to reflect changes in nutritional status of the mothers and could be depressed by nutritional supplementation. Supplementation during pregnancy as well as lactation produced an even greater effect, shortening both the period of post-partum amenorrhoea and mean interbirth interval (Lunn, Austin & Whitehead, 1984). Prema *et al.* (1981), citing difficulties of interpretation in studies involving seasonal variations and nutritional interventions, reported a strong correlation between body weight and the duration of lactational amenorrhoea among poor Indian women after controlling for the length of lactation ($r = -0.81$). In an interesting three-way comparison of Yolngu Aboriginal women of Arnhem Land, Australia, with Konner's !Kung San sample and La Leche League mothers from Boston (Elias *et al.*, 1986), Rich found that the Yolngu were closer to the more rapid Boston pattern of resumption of menses despite being almost indistinguishable from the !Kung in the pattern of their breastfeeding (Rich, 1984). The much heavier body weights of the Yolngu women compared with the !Kung San may be a contributing factor.

Opposition to the idea of a demographically significant interaction between nutritional status and lactational subfecundity has been strong, however. Bongaarts (1980), one of the original authors of the Delgado study, has argued that whatever effect is attributable to nutritional status pales beside the effect of lactation. Jelliffe & Jelliffe (1981) severely criticized the methodology of Lunn's studies, citing inadequate control of nursing variables and nutritional intake. Wood (1985) has taken issue with the statistical significance of Lunn's results.

It may well be that a complete understanding of the relationship of lactational subfecundity and other constitutional and ecological variables, such as age, disease burden, workload and nutritional status, will be achieved only after the individual effects of each of those variables have been clarified. Yet even at this stage, the evolutionary logic of lactational subfecundity can be appreciated (Short, 1984). While a mother's nutritional reserves and intake may be sufficient to meet, for the most part, her own maintenance requirements as well as the requirements of her nursing child (Prentice & Whitehead, 1987), they are rarely sufficient to cover adequately the requirements of a gestating foetus as well. Several studies based on WFS data have now documented strong relationships between birth spacing and child survival (Cleland & Sathar, 1984; Hobcraft, McDonald & Rutstein, 1985; Palloni & Millman, 1986, but see Potter, 1987). In addition to the effect on child survival, short birth intervals may contribute to declining health and nutritional status of mothers (Jelliffe & Jelliffe, 1978). These considerations make it easy to believe that the fitness of individual women living under formative ecological circumstances would ultimately benefit from broadly spaced birth intervals.

Energetics and ovarian function

The interaction of energetics and female reproductive function is currently an area of intense research interest. The Frisch hypotheses originally drew attention to weight and fatness as indicators of female nutritional status and suggested their role in modulating ovarian function (Frisch & Revelle, 1970; Frisch & McArthur, 1974; Frisch 1987). Particularly supportive of this view were the results of studies of anorexia nervosa patients and severely underweight women (Espinos-Compos *et al.*, 1974; Frisch & McArthur, 1974; Warren *et al.*, 1975; Vigersky *et al.*, 1977; Nillus & Wide, 1979), where evidence points to impaired release of luteinizing hormone (LH) as a source of the dysfunction. Despite the fact that anorexia nervosa is complicated by the presence of psychiatric symptoms (Scott & Johnston, 1985) and the lack of complete concordance between weight histories and menstrual histories (Fishman, 1980), the resumption of cycling in response to simple refeeding and weight gain (McArthur *et al.*, 1976) is compelling evidence for the direct effect of body weight and/or composition on ovarian function, at least at this extreme level. Confirming evidence comes as well from studies based on weight loss due to voluntary dieting, rather than psychiatric pathology. Warren *et al.* (1975) and Vigersky *et al.* (1977) both found impaired response to luteinizing hormone-releasing hormone (LHRH) in dieting women based on their degree of weight loss. Warren and her colleagues found that severely underweight women (more than 25 % below Metropolitan Life standards) showed significantly lower responses to LHRH than women classified simply as 'underweight' (15–25 % below standard weight) or normal weight (less than 15 % below standard weight). Luteinizing hormone response increased exponentially at approximately 15 % under standard weight. Similarly, Vigersky *et al.* found delayed LH response to LHRH administration in 19 patients with amenorrhoea secondary to simple weight loss, with a significant correlation between the delay in the LH response and the percentage below standard weight. In a recent study comparing 376 women seeking treatment for infertility of unknown cause with age-matched controls who had successfully conceived at the time that the study subjects had been trying to become pregnant, Green, Weiss & Daling (1988) report that a body weight less than 85 % of Metropolitan Life standards is associated with 4.7-fold increase in the risk of infertility among nulligravid women.

More moderate degrees of underweight may be associated with impairment of ovarian function short of amenorrhoea, such as ovulatory failure and/or luteal phase suppression. Bates & Whitworth (1982) report evidence of shortened luteal phases based on basal body temperature recordings in women being evaluated for infertility who had voluntarily reduced to 91 % of standard weight on average. The women showed reduced LH:FSH ratios which correlated with their degree of underweight. When weight was restored to within 5 % of normal, 73 % of the women conceived spontaneously. These observations suggest that effects on fertility are not confined to extreme cases of undernutrition (Bates, 1985).

Other evidence suggests that energy balance may have an effect on ovarian function independently of nutritional status. Lev-Ran (1974) and Graham, Grimes & Campbell (1979) both document amenorrhoea associated with precipitous weight loss in overweight and normal weight women. Pirke *et al.* (1985, 1988) have found evidence of impaired luteal function and ovulatory failure in women of normal weight-for-height while on rather severe calorie-restriction diets of 800–1000 kcal/day. Catherine Lager and I compared the salivary progesterone profiles of 18 cycles from dieting women who were losing 1.9 kg/month on average with 19 cycles from age-matched controls of stable weight, all the subjects being of normal weight-for-height both before and after the study (Lager & Ellison, 1987; Lager & Ellison, 1990). The dieters had significantly lower levels of average luteal progesterone (214 ± 23 pmol/litre vs. 287 ± 30 pmol/litre, $p < 0.005$) and peak progesterone (461 ± 67 pmol/litre vs. 665 ± 46 pmol/litre, $p < 0.005$) compared with the controls. Only 62 % of the dieters' cycles showed significant evidence of ovulation compared with 100 % of the control cycles. In addition, there was evidence that progesterone profiles were more profoundly suppressed during the month following that in which weight loss occurred, even if weight was stable or increasing at that time, indicating a time-lag in the full effect of weight loss on ovarian function. Studies we have conducted among Efe pygmies and Lese horticulturalists in the Ituri Forest of Zaire (Ellison, Peacock & Lager, 1986, 1989b) indicate that comparable effects may be observed when weight loss is not a matter of fashion but the product of a perilous subsistence economy. Among 35 Lese women, for example, who lost an average of 0.95 kg over a 4-month period in 1984 due to moderate food shortage, ovulatory frequency declined almost linearly with time from 63 % at the beginning to 30 % at the end of the study.

Aerobic exercise has also been shown to be associated with suppression of ovarian function. High prevalence of amenorrhoea and oligomenorrhoea has been noted among ballet dancers (Frisch, Wyshak & Vincent, 1980a; Warren, 1980) and female athletes, especially runners, swimmers and gymnasts (Frisch et al., 1980; Warren, 1980; Dale, Gerlach & Wilhite, 1979; Schwartz et al. 1981) with good evidence that changes in hypothalamic function are responsible (Cumming et al., 1985a,b; Veldhuis et al. 1985). Shangold et al. (1979), in a study of a single female runner, found that training during the luteal phase of her cycle led to a reduction in luteal progesterone, compared with a control cycle. Further, a longitudinal record of luteal phase length based on cervical mucus changes indicated that luteal phase length declined linearly with weekly training mileage, even at very moderate levels.

Trained athletes may not, however, provide an unbiased example of female reproductive function (Malina, 1983). They may, for example, be self-selected for aspects of physique and physiology which predispose them toward lower levels of ovarian function. In addition, it is often difficult to distinguish the effects of exercise from those of nutritional status. Bullen et al. (1985) addressed these problems by implementing an experimental design in which 28 previously untrained women were subjected to a 2-month regime of heavy aerobic exercise. Half the subjects were allowed to lose weight during the training period, the other half received caloric supplements to their diet sufficient to maintain their body weight despite the training. Only four of the subjects failed to show evidence of disrupted ovarian function during the study. Abnormalities, including shortened or deficient luteal phases and failure of the LH surge, were more severe in the weight loss group than the weight maintenance group, and more severe the second month than the first.

Less extreme levels of exercise have also been associated with suppression of ovarian function. Lager and I studied salivary progesterone profiles in a group of recreational women runners who averaged only 20 km a week, all of whom were of normal weight-for-height and menstruating regularly, comparing them with cycles from age-matched sedentary controls (Ellison & Lager, 1985, 1986). Runners had lower and shorter profiles of luteal progesterone than the controls, despite having cycles of similar length and being slightly heavier for their height. This effect may be mediated by reduced frequency and amplitude of LH pulses, as reported by Cumming *et al.* (1985*b*) for eumenorrhoeic subjects who ran for at least 32 km per week.

The potential impact of exercise on fertility has been documented by Green *et al.* (1986) in a retrospective study of 346 infertile women. Exercise histories of the subjects during the period of their unsuccessful efforts to conceive were compared with those of age-matched subjects who had successfully conceived during the same period. Vigorous exercise for an hour or more per day was more prevalent in the infertile group. Among nulligravid women without additional evidence of tubal dysfunction, exercise at this level was associated with a sixfold greater risk of infertility compared with less than an hour of exercise a day.

These results all support the notion that reproductive responses to exercise may be functional rather than pathological (Prior, 1985), and suggest that an argument based on evolutionary logic can be made similar to Frisch's argument concerning body composition; that is, that excessive energy expenditure, especially if chronic and/or compounded with weight loss, represents an increase in the risk of reproductive failure, with suppression of ovarian function being an adaptive response. Bentley (1985) has applied this logic to a reanalysis of !Kung San data, noting its particular relevance to the life-style of these nomadic hunter–gatherers. What are missing, however, are studies of traditional women's workloads and their effects on ovarian function, since it is not clear whether aerobic exercise in a western context provides a useful analogy for these purposes.

Age and ovarian function

Age variation in menstrual function has long been recognized (Collett, Wertenberger & Fiske, 1954), though its relationship to fertility has not always been appreciated. Doring (1969) first published a comprehensive study of ovarian function relative to age on the basis of basal body temperature recordings from 3264 cycles. His results indicated that ovulatory frequency and luteal sufficiency increase steadily from menarche until the middle of the third decade and begin to decline again after the age of 35 years. The relationship of this age-specific curve of ovarian function to age-specific curves of female fecundity and fertility is a matter of intense current interest. At least three segments of this curve deserve particular attention: the initiation of ovarian function, its progressive maturation to peak function, and the phase of decline at older ages.

The exact mechanisms determining the onset of ovarian maturation remain obscure despite concentrated scrutiny. What does seem true is that the maturation of ovarian function is closely tied to the final phases of physical growth. Key events include the slow erosion of hypothalamic sensitivity to negative feedback from circulating oestrogens, and the appearance and facilitation of a hypothalamic 'positive feedback response' capable of triggering an LH surge in response to high levels of oestradiol. Experiments conducted by Knobil and his associates in rhesus monkeys (Knobil et al., 1980; Wildt, Marshall & Knobil, 1980) indicate that these effects can be produced artificially by exposing the pituitary to hypothalamic gonadotrophin-releasing hormone (GnRH) in a regime of hourly pulses mimicking the adult pattern, while in humans the same maturational effects can be averted by artificially overriding a quite similar GnRH release pattern (Crowley et al., 1985). The implication is that an adult GnRH release pattern is both necessary and sufficient to induce these maturational changes, and that there is no constraint associated with the possible immaturity of the pituitary, ovary, or uterus which might affect the onset of the process of maturation. Multiple hypotheses have been offered to account for the timely maturation of the hypothalamus, including a simple endogenous schedule of maturation (Grumbach et al., 1974; Ojeda, Advis & Andrews, 1980), exposure to rising titres of adrenal steroids and their conversion products (Parker & Mahesh, 1976, 1977; Sizonenko, 1978), and physiological signals associated with body composition (Frisch, 1980).

Frisch has argued strongly that the logic of energetic constraints on reproductive success dictates that the timing of reproductive maturation should be tightly linked to accumulation of sufficient energy reserves to support successful reproduction, in the form of body fat, whatever the proximate mechanism, and a substantial body of data can be lined up consistent with this view. Indeed, several of the proposed mechanisms could operate to this end, including those based on the aromatization of adrenal androgens in adipose tissue (Frisch, Canick & Tulchinsky, 1980b; Ellison, 1984), and the direction of oestrogen metabolism (Fishman, Boyar & Hellman, 1975).

I have also suggested, however, that there are mechanical constraints on reproductive success logically prior to energetic constraints, especially those associated with pelvic size, that until a women is 'big' enough to reproduce the question of whether she is 'fat' enough is moot (Ellison, 1982). By this logic the essential linkage is between reproductive maturation and the attainment of an appropriate physical size for reproduction. Again, data support the idea that the constraints of pelvic size on reproductive success are significant (Greulich & Thoms, 1939; Baird, 1952; Bressler, 1962; Thompson & Billewicz, 1963; Liestol, 1980; Wyshak, 1983), and that the cessation of physical growth and onset of reproductive maturation are timed to result in similar pelvic dimensions, whether that be early or late (Ellison, 1982; Moerman 1982). The association of menarcheal age with measures of skeletal and pelvic maturation is greater than that with weight or fatness (Ellison, 1981a, 1982), and holds both within and between populations (Ellison, 1982).

Once ovarian cycles are initiated, their maturation continues progressively until the early to mid-twenties. Studies of plasma hormonal profiles have shown a high frequency of anovulatory and luteally insufficient cycles in the first year after menarche, with a progressively rising trajectory of ovulatory frequency and luteal competence in the succeeding years (Apter, Viinkka & Vihko, 1978; Vihko & Apter 1984; Venturoli *et al.*, 1987). At least one study based on urinary pregnanoediol determinations has confirmed this pattern of ovarian maturation in the years after menarche (Metcalf *et al.*, 1983). Read *et al.* reported on a study of salivary progesterone levels which also provided evidence of low ovulatory frequency among post-menarcheal girls aged 12 to 17 years (Read *et al.*, 1984). This study also found that the progesterone levels during the luteal phases of ovulatory cycles were lower among women in this age range than among older women. We found significantly lower salivary progesterone profiles and lower ovulatory frequency among women aged 18 to 22 years than among 23- to 35-year-old women of similar height and weight, even though menstruation was well established in these subjects (Ellison, Lager & Calfee, 1987). Further study of the ultradian pattern of episodic progesterone in saliva collected every 15 minutes has revealed a distinctly different pattern of progesterone release in the luteal phases of 19- to 21-year-old women compared with 25- to 35-year-old women, pointing to a diminished capacity for secreting progesterone on the part of the corpus luteum in younger women (cf. below, pp. 35–6). Many of these differences in ovarian function would not be suspected on the basis on menstrual histories.

The decline in ovarian function with increasing age after 35 years is less well characterized. If analysis is restricted to women who continue to experience regular menstrual cycles, ovulatory frequency appears to decline only gradually between 35 and 45 years among German women, Boston women and Lese women from Zaire (Doring, 1969; Ellison *et al.*, 1989b; Fig. 2.2). There is, however, some evidence to suggest that luteal suppression becomes more prevalent in this interval. Doring found an increased incidence of shortened luteal phases based on basal body

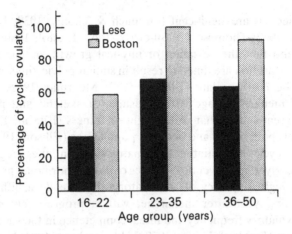

Fig. 2.2. Distribution of ovulatory frequency by age among Lese and Boston women (Ellison *et al.*, 1989*b*).

temperature records, especially among women of 46 to 50 years old (Doring, 1969). Preliminary results from an ongoing study of age-related changes in ovarian function based on salivary steriod analyses being carried out by my research group also suggests that luteal function is suppressed in women over the age of 40 years relative to women in their twenties and thirties. The average progesterone profile obtained for women 40 to 44 years of age is comparable in shape and level to that obtained for women 18 to 19 years of age (Fig. 2.3).

Attempts to quantify directly the effect of increasing age on fecundity have yielded suggestive results. A controlled study of the success rate of artificial insemination by donor as a function of age in a large sample of

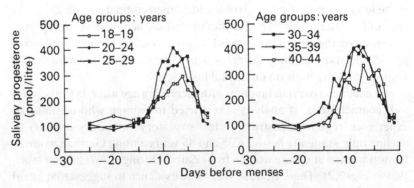

Fig. 2.3. Average salivary progesterone profiles for six age groups of Boston women ($N = 90$).

French women found fecundity to decline significantly after the age of 35 years with the decline accelerating after age 40 years (Federation CECOS, Schwartz & Mayaux, 1982). Menken, Trussell & Larsen (1986) have analysed historical material from the US to identify age-related changes in female fecundity after correcting for factors such as duration of marriage and age of spouse. They found a pattern of declining fecundity which, though present in 35- to 39-year-old women, was not striking until after the age of 40 years. They were also able to document the historical stability of this pattern, suggesting that such age-related changes in fecundity are not a unique pathology of our contemporary world, but are rather normal features of our reproductive biology.

Particularly noteworthy have been studies which suggest that the trajectory of ovarian maturation and the ultimate adult level of ovarian function may vary with overall developmental tempo. The principal evidence for this comes from the work of Apter & Vihko (1983; Vihko & Apter, 1984). In studying the menstrual cycles of adolescent girls, they found that ovarian maturation proceeded at an absolutely faster rate among girls with an earlier menarcheal age than among those with a later menarcheal age. Girls who reached menarche at or before age 12 years, for example, achieved a 50% ovulatory rate within 1 year after menarche, while girls with a menarcheal age of between 12 and 13 years took 3 years, and those with a menarcheal age over 13 years took 4.5 years to reach the same ovulatory frequency. Based on a separate longitudinal study, Venturoli *et al.* (1987) suggests that early onset of regular cycles leads to an earlier establishment of adult endocrine profiles of ovarian function; persistent irregular, but ovulatory, cycles lead to a slower trajectory of ovarian maturation which eventually catches up with that of the early maturers; persistent anovulatory and irregular cycles possibly represent a basic defect that often continues into adulthood. At least some data, including Apter & Vihko's original study (1983), suggest that the trajectory of ovarian function followed by the late maturers may never reach the same level of ovulatory frequency that early maturers achieve. To this may be added data from studies of linkages within the gynaecological histories of individual women (Gardner, 1983; Gardner & Valadian, 1983) which indicate that late maturers continue to have higher incidences of oligomenorrhoea and dysmenorrhoea than early maturers well into their reproductive prime.

A good body of cross-cultural data does not yet exist to test whether the relationship between tempo of growth and trajectory of ovarian function also holds between populations. It is perhaps noteworthy that among the Lese of Zaire, where menarche does not ordinarily occur until roughly age 16 years, the trajectory of ovulatory frequency by age parallels that found among Boston women, but at a significantly lower level (Fig. 2.2). Nor can that difference in level be accounted for entirely by the suppressive effects of weight loss and nutritional stress (Ellison et al., 1989b). Should this relationship hold up between other populations, an important corollary would follow: adolescent subfecundity should be much more significant in limiting the fertility of young women in populations with late, as opposed to early, physical maturation, even comparing women of comparable gynaecological age. In a culture that ranks teenage pregnancy among its most pressing social concerns, we may be too ready to discount adolescent subfecundity as a phenomenon of importance.

Disease and ovarian function

Relative ignorance of the relationship between chronic disease burden, other than venereal disease, and reproductive function represents perhaps the largest gap in our understanding of human reproductive ecology. Although it is clear that serious and debilitating illnesses can disrupt normal ovarian activity (McFalls & McFalls, 1984), comparatively little is known about the effects of more moderate levels of morbidity. The acute, spiking fevers of malaria have been associated with elevated rates of early embryonic loss (Lewis, Lauersen & Binnbaum, 1973; Morishima et al., 1975), though probably more as a consequence of effects on uterine contractility than as a consequence of changes in ovarian function.

Veneral diseases such as syphilis, gonorrhoea, and chlamydia are clearly capable of adversely affecting the fertility of individuals, and even of populations. The African 'infertility belt', a region straddling the equator characterized by isolated pockets of dramatically low fertility and high rates of secondary sterility, has long been thought to be a demographic representation of epidemic outbreaks of these diseases (Belsey, 1976; Caldwell & Caldwell, 1983; Frank, 1983). While tubal blockage is the ordinary cause of infertility secondary to these diseases, ovarian involvement may occasionally occur leading to abnormal ovarian function (Toth, Senterfit & Ledger, 1983).

Much more evidence exists on the general relationship between ovarian and immune function. Luteal progesterone production, for example, has been shown to have an inverse relationship to the rate of *Chlamydia trachomatis* inclusions *in vitro* and to the degree of resulting inflammation and productive infection *in vivo* (Tau-Cody et al., 1988). An inverse relationship has also been demonstrated between ovarian progesterone and the production of interleukin-I (IL-1), a monokine mediator of immune response (Polan, Carding & Loukides, 1988). IL-1 itself acts at the level of the hypothalamus to stimulate release of corticotropin-releasing factor (CRF) (Woloski *et al.*, 1985; Besedovsky *et al.*, 1986; Berkenbosch *et al.*, 1987; Bernton *et al.*, 1987; Sapolsky *et al.*, 1987; Breder, Dinarello & Saper, 1988). CRF in turn may act centrally to inhibit reproductive function (Rivier, Rivier & Vale, 1986). Some evidence also suggests that steroids, especially progesterone, may be natural ligands of σ receptors on peripheral human lymphocytes, serving to further mediate immune function (Su, London & Jaffe, 1988). Relationships such as these may account in part for the chronically low levels of luteal function, independent of other ecological variables, among Lese women in Zaire.

Male factors

Although implicated in many cases of individual infertility, male factors have rarely been associated with depressed fertility on a population level. This is not because the male reproductive system is impervious to environmental variables. Severe malnutrition has long been known to suppress male testicular function (Keys *et al.*, 1950; Zubiran & Gomez-Mont, 1953; Gomez-Mont, 1959), though the pathway may be different for chronic and acute undernutrition. Chronic severe malnutrition appears to result in clinical hypogonadism, typified by reduced serum testosterone concentrations, elevated gonadotropins, and decreased responsiveness to exogenous gonadotropin administration (Gomez-Mont, 1959; Smith *et al.*, 1975), all suggestive of a defect at the level of the testis. Alternatively, weight loss during fasting is associated with reductions in both testosterone and gonadotropins (Lee *et al.*, 1977; Klibanski *et al.*, 1981), suggesting alterations in hypothalamic–pituitary function.

Most work has concentrated on the effects of either acute fasting or severe chronic undernutrition. We studied salivary testosterone levels in Efe and Lese males from the Ituri Forest of Zaire (Ellison, Lipson & Meredith, 1989*a*), where our longitudinal data indicate nutritional intake is normally low but adequate, punctuated by seasonal shortfalls when significant weight may be lost over a period of months (Bailey & Peacock 1988; Bailey & DeVore 1989; Dietz *et al.*, 1989). Although our results indicate testosterone levels in the Ituri males are individually within the broad normal ranges established for western populations, the average values for both Efe and Lese males are significantly below those of age matched Boston males. Although an association has yet to be established between testosterone levels and recent nutritional history for the Ituri

34   P. T. Ellison

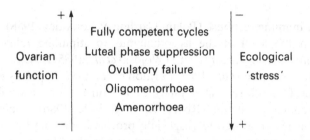

Fig. 2.4. Graded continuum of ovarian response to environmental stress.

males, a significant positive relationship between salivary testosterone and adult height among the Lese suggests that chronic and/or constitutional factors may be important (Ellison *et al.*, 1989*a*).

Aerobic exercise also has clear effects on male reproductive function (Wall & Cumming, 1985). The acute effect of exercise is often an increase in circulating testosterone (Cumming *et al.*, 1986), though this is at least in part a function of lower clearance rates resulting from decreased hepatic blood flow. Chronic exercise effects, however, are usually in the opposite direction, leading to decreased testosterone (Wheeler *et al.*, 1984), presumably resulting from changes in hypothalamic–pituitary function (MacConnie *et al.*, 1986).

Ovarian function and the regulation of human fertility

Although the field of human reproductive ecology is far from static, and although new results continue to appear daily, it is possible to generate some new, middle level hypotheses on the role of ovarian function in regulating human fertility to help integrate our current state of knowledge while articulating explicitly the relationship of these accumulated observations to broader evolutionary theory.

1. Ovarian function shows a graded response to constitutional, environmental, and behavioural variables. Early studies of age, lactation, nutritional and exercise effects on female reproductive physiology stressed the onset and maintenance of regular menstrual cycles. It is now apparent that disruptions in menstrual function represent the extreme end of a graded continuum of ovarian response to these variables (Fig. 2.4). While rapid or extreme weight loss (Graham *et al.*, 1979; Pirke *et al.*, 1985), heavy exercise (Bullen *et al.*, 1985), early adolescence (Apter & Vihko, 1983) and frequent, intense nursing characteristic of the early post-partum period (Howie *et al.*, 1982; McNeilly *et al*, 1982) are all associated with irregular or absent menstrual cycles, moderate weight loss (Bates *et al.*, 1982; Lager & Ellison, 1987; Lager & Ellison, 1990), moderate exercise (Ellison & Lager, 1985, 1986), late adolescence (Ellison *et al.*, 1987) and less frequent nursing characteristic of the period of introduced solid foods (Howie *et al.*, 1981, 1982) are all associated with reduced ovulatory frequency and luteal suppression, even though menses may be regular. Movement across the gradient of responses can be observed in all these cases, toward diminished ovarian function with progressive weight loss (Ellison *et al.*, 1989*b*), extended exercise (Bullen *et al.*, 1985), or increasing age after 35 years (Doring, 1969), and toward augmented ovarian function with increasing age during adolescence (Apter & Vihko, 1983; Read *et al.*, 1984; Ellison *et al.*, 1987), weight gain or stabilization (Lager & Ellison, 1990) or declining nursing intensity post-partum (Howie *et al.*, 1981, 1982).

Ordinarily, only the extreme end of this continuum lies above the clinical horizon. Detection of moderate suppression of ovarian function requires more sophisticated and often more invasive procedures, and conditions in this range are likely to go unnoticed by both a woman and her physician. Moderate suppression of ovarian function may, nevertheless, have significant effects on a woman's fecundity (Bates, 1985; Green *et al.*, 1986; Green *et al.*, 1988). Ovulatory failure, which may be associated with follicular phase suppression of follicle growth and oestradiol production (Stanger & Yovich, 1984; Apter *et al.*, 1987; Eissa *et al.*, 1987) or with disruption of the functioning of the hypothalamic–pituitary axis (Cumming *et al.*, 1985*a,b*; Veldhuis *et al.*, 1985), clearly reduces the probability of conception to zero for that month. Yet milder forms of follicular suppression, without precluding ovulation, may lead to an ovum of reduced fertilizability (Yoshimura & Wallach, 1987). Similarly, luteal phase suppression of progesterone production diminishes the likelihood of successful implantation and increases the probability of early pregnancy wastage (Tho, Byrd & McDonough, 1979; Rosenberg, Luciano & Riddick, 1980; McNeeley & Soules, 1988).

Recent studies of ultradian patterns of ovarian steroid production suggest additional subtle manifestations of mild ovarian suppression (Hughes *et al.*, 1987; McNeely & Soules, 1988; O'Rourke & Ellison, 1990). The late luteal phase of mature women, for example, is characterized by progesterone pulses of high amplitude but low frequency (O'Rourke, 1987; O'Rourke & Ellison, 1990). Such transient surges of progesterone may be crucial in what would be the peri-implantation period of a conception cycle. There is even evidence that the hypothalamus may contain two separate progesterone receptor systems, one based

Table 2.1. *Summary of pulse characteristics by age group and cycle phase among Boston women (N = 17; mean ± SE; Thorne, 1988)*

Age group (years)	Mid-luteal phase			Late luteal phase		
	Frequency (pulses/h)	Amplitude (pmol/litre)	Duration (h)	Frequency (pulses/h)	Amplitude (pmol/litre)	Duration (h)
19 to 21	0.50 ± 0.30	200 ± 69	0.25 ± 0.0	1.75 ± 0.48	102 ± 17	0.47 ± 0.15
24 to 34	3.75 ± 0.85	139 ± 19	0.82 ± 0.20	1.80 ± 0.74	208 ± 28	0.40 ± 0.12

on membrane receptors responding to acute changes in progesterone levels, the other based on cytosol receptors and responding to tonic progesterone levels (Ramirez, Kim & Dluzen, 1985). Differences in pulsatile patterns do exist associated with at least one of the variables considered here. Preliminary data from our laboratory suggest that young women (19–21 years old) have a significantly lower frequency of progesterone pulses than older women (23–34 years old) in the mid-luteal phase and significantly lower amplitude in the late luteal (Table 2.1). Yet the detection of progesterone pulse patterns is only possible from samples drawn at frequent intervals for an extended period.

The key feature of the continuum of ovarian response is its graded character. That is, ovarian physiology has built into it critical transitions which represent abrupt, nearly discontinuous changes in the probability of reproductive success. As noted, follicular phase suppression of oestradiol production may lead to an ovum of diminished fertilizability (Yoshimura & Wallach, 1987) and possibly to a corpus luteum of reduced capacity and hence a lower probability of implantation (McNeely & Soules, 1988), but at some point a threshold is passed resulting in ovulatory failure and a probability of conception for that month of zero. More profound suppression of ovarian function, leading to oligomenorrhoea and amenorrhoea represent lengthier periods of zero fecundability. This graded nature of the ovarian response continuum provides a mechanism for 'stepping down' female fecundity in stages in response to ecological variables which may be changing or cumulating smoothly.

2. *The graded continuum of ovarian response forms a 'final common pathway' through which various ecological variables affect female fecundity.* As noted above, weight loss, exercise, youth, and lactation all share the graded continuum of ovarian suppression as a final

common pathway through which they modulate female fecundity. The mechanisms through which they enter this pathway, however, may well be diverse. Elevated prolactin levels associated with lactation have been shown both to alter hypothalamic regulation of pituitary gonadotropin release (Baird *et al.*, 1979) and to inhibit gonadal steroid production (McNatty *et al.*, 1979). Intense exercise disrupts the hypothalamic–pituitary axis (Cumming *et al.*, 1985*a,b*), possibly through increased levels of endogenous opioids (Knobil *et al.*, 1980; Howlett *et al.*, 1984; Yen, 1984), though testosterone and prolactin levels are raised acutely as well (Shangold, Gatz & Thyssen, 1981). Weight loss is associated with decreased production of extragonadal oestrogens (Siiteri, Williams & Takaki, 1976; Takaki *et al.*, 1978; Pinter *et al.*, 1980; Ellison, 1984) and hypothalamic dysfunction (Warren *et al.*, 1975; Vigersky *et al.*, 1977). Adolescent cycles are characterized by elevated androgen to oestrogen ratios (Apter *et al.*, 1978), strong circadian variations in pituitary activity (Boyar *et al.*, 1974), and reduced capacity for steroid production (Ellison, O'Rourke & Thorne, in prep.).

The existence of a final common pathway of ovarian suppression is an asset to some questions and applications, since ovarian activity is much easier to monitor under field conditions and for extended periods than the activity of the hypothalamus or pituitary (Ellison, 1988). Changes 'above' the level of the ovary in the HPO axis express themselves functionally in changes in the production of ova and their quality, while many important events 'below' the level of the ovary, such as implantation and early gestation, are guided and regulated by ovarian steroid production. Ovarian function thus becomes the 'lynchpin' of a complex, multilevel physiological system, with steroid levels being the most useful reflection of ovarian function. The negative consequence, however, is that ecological studies of ovarian function may be unable to distinguish between competing causes of ovarian suppression. Attention to ultradian patterns of steroid production may possibly resolve this problem by distinguishing, for instance, between luteal suppression resulting from inhibition of steroid production (diminished progesterone pulse amplitude and/or duration) and interference with gonadotropin release (diminished pulse frequency).

3. *Ovarian function discriminates in its response between eco-logical 'noise' on the one hand, and significant changes or ecological 'trends' on the other.* To presume that human reproductive physiology has been moulded by natural selection is to presume that the ovarian responses I have detailed ordinarily benefit the fitness of individual

women, or would have done so under formative ecological and social conditions. It has been noted that it may appear paradoxical for reproductive failure *ever* to be classified as an adaptive response, since evolutionary fitness is measured in units of reproductive success (Peacock, 1990). Yet the ovarian responses I have noted should really be considered as mechanisms to *conserve* energy and resources when the prospects of reproductive success are low, limiting reproductive effort as early as possible in the temporal sequence of fertilization, implantation, gestation, parturition and lactation. The result is perhaps more easily understood as a reduction in reproductive success *variance*.

Environmental tracking is the key feature of short-term ecological regulation of ovarian function. While severe stresses may curtail ovarian function abruptly, the graded continuum of ovarian response operates so that mild stresses tend to set the stage for early pregnancy wastage without irrevocably committing things to that course. The woman's physiology is placed 'on alert' for a continuation of unfavourable conditions that may result in a more serious condition or betoken a more long-lasting trend. Weight loss, for example, which continues for several months may be 'interpreted' as evidence of unfavourable nutritional conditions with fecundity lowered in response, even if the current nutritional status of the woman is good. Similarly, a consistent period of several months' weight gain may be construed as evidence of an ameliorating nutritional environment favourable to reproduction with fecundity raised in response, even if a woman has not yet recovered her 'ideal' nutritional status. Physiological responses which seem pathological in the context of often highly idiosyncratic and episodic histories of self-imposed weight loss in western women may in fact be adaptive responses in the context of formative human environmental conditions.

More severe environmental stresses lead to a transition of effect, from an enhanced likelihood of early pregnancy wastage to a reduced likelihood of pregnancy itself. The crucial feature of the graded ovarian response is that it carries with it inherent transitions of this kind between levels of fecundity. One of the persistent difficulties with the Frisch hypothesis has been the problem of *how* the reproductive system could recognize critical thresholds in body composition in order to respond with a dramatic qualitative change in function. It now appears that the natural transition points in ovarian function may provide for such transitions in fecundity. That is, while ovarian function appears to change rather smoothly in response to environmental and constitutional factors, the innate transitions in state – from luteal competence to luteal insufficiency, from ovulatory to anovulatory – may translate into a relationship between such factors and fecundity that more closely resembles a step function. From this perspective, ovarian function plays a buffering role between environmental perturbations and fertility. As Prentice & Whitehead have noted (Prentice & Whitehead, 1987), the low energy cost of human reproduction per unit time relative to other mammals together with a high capacity for energy storage enable our species to make effective use of buffering strategies. The graded response of the ovary serves to distinguish mild and short-term environmental perturbations from drastic changes and longer-term trends, sustaining fecundity in the former case and adjusting it in the latter. Particularly important for improving our understanding in this area will be longitudinal studies of ovarian function focusing on physiological linkages between cycles which might help clarify the way in which environmental effects are cumulated.

4. Developmental tracking of ovarian function serves to establish relative sensitivity to ecological variables. If we think of the graded continuum of ovarian response as representing a kind of 'dose-response curve' relating environmental 'stresses' to changes in female fecundity, a key question becomes the positioning of that curve relative to the environmental axis. That is, for example, at what rate of weight loss or what level of exertion should ovulation be interrupted? How long a period of weight stabilization or how much weight gain should occur for menses to resume in a woman recovering from starvation amenorrhoea? It is clear that variability exists between women in both level and responsiveness of ovarian function; what we need to understand is how that variability is organized.

An important clue is provided by the relationship between overall developmental tempo and the trajectory of ovarian function by age. Slow physical development is most often a consequence of unfavourable nutritional or epidemiological conditions, or of a constitution with higher than normal energy requirements. If ovarian function is truly sensitive to these same variables, then it would make sense for thresholds of ovarian sensitivity to be set, at least in part, by the pace of physical development. That is, to the degree that an environment can be characterized as chronically 'hostile', a woman may best serve her own fitness by being highly conservative in her allocation of reproductive effort and forgoing marginal opportunities for reproduction which others in chronically more favourable settings might undertake. Mechanisms such as these could account for the chronically lower level of ovarian function among Lese women compared with Boston women, the result not being so much chronically reduced fertility as heightened sensitivity to ecological variables such as weight loss. This hypothesis would also predict that we should find wide population variability in ovarian function even as we do in adult height and the tempo of growth and development. We await the accumulation of a sufficient base of cross-cultural data to test this prediction.

As work proceeds on all these fronts, we must keep in mind the lesson of research on lactational subfecundity, that relationships which seem vague or weak may suddenly appear clearer and stronger when the 'right' variables are used in the analysis. Similarly, interaction effects among the variables considered here, which may very well bear the most realistic relationship to human experience, will most likely not be understood properly until we achieve a firm grasp of main effects.

Reproductive ecology and human affairs

I wish briefly to comment on two areas in which the study of human reproductive ecology, particularly from the perspective of biological anthropology, intersects practical matters of health and policy. The first concerns the fresh perspective that this discipline can bring to the treatment of individual infertility and the interpretation of syndromes such as luteal phase deficiency. The second concerns the proper contribution of reproductive ecology to the formulation of public policy in areas such as maternal and child health and population policy.

Inadequate functioning of the corpus luteum in the second half of the menstrual cycle is a common manifestation of female infertility patients without lesions of the reproductive tract, and is especially prevalent among women suffering from recurrent abortion (Smart et al., 1982; Huang, Muechler & Bonfiglio, 1984; Maslar, 1988; Soules et al., 1988). Several recent reviews (Lee, 1987; McNeeley & Soules, 1988; Stouffer, 1988) in the literature of reproductive medicine have struggled with the problem of developing good criteria for clinically diagnosing 'luteal phase deficiency' (LPD). In all of these reviews the standard diagnostic approaches – basal body temperature recordings, ultrasonography, single serum progesterone determinations, multiple serum progesterone determinations, endometrial biopsy – are all criticized for failing to produce a clean separation of LPD patients from normal controls. The conclusions are discouraging over the prospect of developing useful criteria for diagnosis, and the frustration of the authors is clear.

From the perspective of a biological anthropologist studying human reproductive ecology, the problem is an artifact of the implicit model of

LPD that the diagnostician has in his head. According to this model, LPD is a pathological state distinct from the healthy functioning of the HPO axis, a state which will reveal itself as a second mode in a distribution plotted on the appropriate axis (Polansky & Lamb, 1989). The perspective of this review suggests a different model, for which the term 'luteal phase suppression' is perhaps more appropriate. That is, rather than a distinct state, inadequate functioning of the corpus luteum is understood as the tail of a single distribution of ovarian function. States such as youth, undernutrition and heavy exercise can suppress luteal function in an individual woman relative to what it would be in their absence, and hence increase the probability of reproductive loss as a consequence. But both luteal function and its effect on fertility are understood as continuous variables. Underlying this model is the basic appreciation for normal human variation that is the intellectual legacy of the biological anthropologist, together with a theoretical orientation which allows for conditions like luteal phase suppression to be understood as healthy, adaptive responses to certain conditions, and not as pathological lesions of normal function. Were it to be accepted by the clinical community, this perspective on human reproduction might well suggest different therapeutic approaches to the management of conditions like luteal phase suppression, highlighting the effect of changes in life-style and de-emphasizing chemical intervention, while at the same time calling off the search for a diagnostic discriminator.

Scott & Johnston (1985) have noted that good health policy can only be based on good science, that prematurely basing policy decisions on hypotheses which remain unconfirmed or controversial, even though they may be very much in vogue, is likely to produce damaging consequences. Their warning was issued specifically with reference to the Frisch hypotheses and the possibility of seeing in them a reason to curtail nutritional supplementation for mothers and infants in the developing world, but its basic wisdom obviously applies much more broadly. I would add, however, that good science, while necessary to the formulation of good policy in areas of human health and biology, is not of itself sufficient, that we should be thoughtful about the policy implications of even the most sound basic research. The proper application of reproductive ecology to matters of public policy, in my mind, is in identifying and anticipating areas of concern, rather than in indicating or legitimizing particular courses of action.

There is at present broad consensus among public health officials that the practice of breastfeeding needs to be encouraged in the developing world (Buchanan, 1975; Rosa, 1976; Population Reports, 1981; WHO/

NRC (USA), 1983; Short, 1984; Scott & Johnston, 1985). The contraceptive effect of lactation is only one of the reasons for this consensus in support of breastfeeding; nutritional and immunological benefits for the child are considerable as well, especially in areas where clean water supplies are insufficient (Jelliffe & Jelliffe, 1978). Furthermore, the work of Prentice, Whitehead and their colleagues (Paul, Mulle & Whitehead, 1979; Prentice et al., 1986) and Linda Adair (1984), among others, demonstrates that women are able to maintain both the quantity and quality of milk production even while under considerable nutritional stress themselves through mechanisms that probably include adjustments in their own energy requirements (Prentice & Whitehead, 1987).

Yet at the same time that one may enthusiastically endorse the promotion of breastfeeding as a health policy priority in the developing world, one must be concerned with the practical limitations of this policy. Breastfeeding patterns are among the most vulnerable elements of traditional women's life-styles in transitional societies (Knodel, 1977; Popkin, Bilsborrow & Akin, 1982). Although it may be possible for vigorous efforts to reverse the trend toward shortening the period of intensive lactation in specific populations (Haaga, 1986), wiser policy would be based on an appreciation of the societal pressures working in the opposite direction as women's educational level, workforce participation and domestic roles change with economic development. Particular attention should be paid to situations in which national development shifts women's productive activities away from heavy physical labour, such as subsistence agriculture, to more sedentary occupations. Certainly, the promotion of breastfeeding should not be construed as an alternative to family planning education and services or primary maternal and child health care.

The implications of age-associated changes in ovarian function for public health policy may also deserve special attention. Concern in our own society over the evidence for declining female fecundity after the age of 35 illustrates how important such issues can become when cultural trends lead to delayed childbearing. For industrialized societies, enlightened health policy can anticipate an increased demand for infertility treatment which will accompany these trends (Menken et al., 1986). More important, however, are the changes at the other end of the age spectrum, where secular trends in reproductive and physical development can suddenly make teenage fertility and reproductive behaviour an urgent source of concern in societies whose educational, legal, religious and health care institutions may all be ill-equipped to deal with the phenomenon of a 13-year-old mother. The greatest challenge for policy makers in developing countries is to anticipate the effects of many of these changes occurring together, increases in fecundity, shifts in the age distributions of fertility, and reduction in the practice of breastfeeding. Good science in the area of human reproductive ecology can help to identify these areas of concern, but cannot of itself provide the answers.

References

Adair, L. S. (1984). Marginal intake and maternal adaptation: the case of rural Taiwan. *Current Topics in Nutrition and Disease*, **11**, 33–55.

Apter, D., Raisanen, I., Ylostalo, P. & Vihko, R. (1987). Follicular growth in relation to serum hormonal patterns in adolescents compared with adult menstrual cycles. *Fertility and Sterility*, **47**, 82–8.

Apter, D. & Vihko, R. (1983). Early menarche, a risk factor for breast cancer, indicates early onset of ovulatory cycles. *Journal of Clinical Endocrinology and Metabolism*, **57**, 82–8.

Apter, D. Viinkka, L. & Vihko, R. (1978). Hormonal pattern of adolescent menstrual cycles. *Journal of Clinical Endocrinology and Metabolism*, **47**, 944–54.

Bailey, R. C. & DeVore, I. (1989). Research on the Efe and Lese populations of the Ituri Forest, Zaire. *American Journal of Physical Anthropology*, **78**, 459–71.

Bailey, R. C. & Peacock, N. R. (1988). Efe pygmies of northeastern Zaire: subsistence strategies in the Ituri Forest. In *Uncertainty in Food Supply*, ed. I. deGarine & G. A. Harrison, pp. 88–117. Oxford: Oxford University Press.

Baird, D. (1952). The cause and prevention of difficult labor. *American Journal of Obstetrics and Gynecology*, **63**, 1200–12.

Baird, D. T., McNeilly, A. S., Sawers, R. S. & Sharpe, R. M. (1979). Failure of estrogen-induced discharge of luteinizing hormone in lactating women. *Journal of Clinical Endocrinology and Metabolism*, **49**, 500–6.

Bates, G. W. (1985). Body weight control practice as a cause of infertility. *Clinical Obstetrics and Gynecology*, **28**, 632.

Bates, G. W., Bates, S. R. & Whitworth, N. S. (1982). Reproductive failure in women who practice weight control. *Fertility and Sterility*, **37**, 373.

Belsey, M. A. (1976). The epidemiology of infertility: a review with particular reference to sub-Saharan Africa. *Bulletin of the World Health Organization*, **54**, 319–41.

Bentley, G. R. (1985). Hunter–gatherer energetics and fertility: a reassessment of the !Kung San. *Human Ecology*, **13**, 79–109.

Berkenbosch, F., Oers, J. V., Rey, A. D., Tilders, F. & Besedovsky, H. (1987). Corticotropin-releasing factor-producing neurons in the rat activated by interleukin-1. *Science*, **238**, 524–6.

Bernton, E. W., Beach, J. E., Holaday, J. W., Smallridge, R. C. & Fein, H. G. (1987). Release of multiple hormones by a direct action of interleukin-1 on pituitary cells. *Science*, **238**, 519–21.

Besedovsky, H., Rey, A. D., Sorkin, E. & Dinarello, C. A. (1986). Immuno-regulatory feedback between interleukin-1 and glucocorticoid hormones. *Science*, **233**, 652–4.

Billewicz, W. Z., Fellowes, H. M. & Hytten, C. A. (1976). Comments on the critical metabolism mass and the age of menarche. *Annals of Human Biology*, **3**, 51–9.

Bongaarts, J. (1980). Does malnutrition affect fecundity? A summary of evidence. *Science*, **208**, 564–9.

—— (1982). Malnutrition and fertility. *Science*, **215**, 1273–4.

—— (1985). Nutrition and lactational amenorrhea: a comment on Hodgson. *Medical Anthropology Quarterly*, **16**, 40–1.

Bongaarts, J. & Potter, R. G. (1983). *Fertility, Biology, and Behavior: an analysis of the proximate determinants*. New York: Academic Press.

Boyar, R. M., Katz, J., Finkelstein, J. W., Kapen, S., Weiner, H., Weitzman, E. D. & Hellman, L. (1974). Anorexia nervosa: immaturity of the 24-hour luteinizing hormone secretory pattern. *New England Journal of Medicine*, **291**, 861–5.

Breder, C. D., Dinarello, C. A. & Saper, C. B. (1988). Interleukin-1 immunoreactive innervation of the human hypothalamus. *Science*, **240**, 321–4.

Bressler, J. B. (1962). Maternal height and the prevalence of stillbirths. *American Journal of Physical Anthropology*, **20**, 515–17.

Brown, J. B., Harrison, P. & Smith, M. A. (1985). A study of returning fertility after childbirth and during lactation by measurement of urinary oestrogen and pregnanediol excretion and cervical mucus production. *Journal of Biosocial Science, Supplement* **9**, 5–24.

Buchanan, R. (1975). Breastfeeding: aid to infant health and fertility control. *Population Reports, Series J*, **4**, 49–67.

Bullen, B. A., Skrinar, G. S., Beitins, I. S., von Mering, G., Turnbull, B. A. & McArthur, J. W. (1985). Induction of menstrual disorders by strenuous exercise in untrained women. *New England Journal of Medicine*, **312**, 1349–53.

Caldwell, C. & Caldwell, P. (1983). The demographic evidence for the independence and cause of abnormally low fertility in tropical Africa. *World Health Statistics Quarterly*, **36**, 2–34.

Cameron, N. (1976). Weight and skinfold variation at menarche and the critical body weight hypothesis. *Annals of Human Biology*, **3**, 279–82.

Cleland, J. G. & Sathar, Z. A. (1984). The effect of birth spacing on childhood mortality in Pakistan. *Population Studies*, **38**, 401–18.

Collett, M. E., Wertenberger, G. E. & Fiske, V. M. (1954). The effect of age upon the pattern of the menstrual cycle. *Fertility and Sterility*, **5**, 437–48.

Corsini, C. (1979). Is the fertility reducing effect of lactation really substantial? In *Natural Fertility*, ed. H. Leridon & J. Menken, pp. 197–215. Liege: Ordina.

Crowley, W. F., Filicori, M., Spratt, D. I. & Santoro, N. F. (1985). The physiology of gonadotropin-releasing-hormone (GnRH) secretion in men and women. *Recent Progress in Hormone Research*, **41**, 473–531.

Cumming, D. C., Brusting, L. A. III, Strich, G., Reis, A. L. & Rebar, W. W. (1986). Reproductive hormone increases in response to acute exercise in men. *Medical Science in Sports and Exercise*, **18**, 369–73.

Cumming, D. C., Vickovic, M. M., Wall, S. R. & Fluker, M. R. (1985a). Defects in pulsatile LH release in normally menstruating runners. *Journal of Clinical Endocrinology and Metabolism*, **60**, 810–12.

Cumming, D. C., Vickovic, M. M., Wall, S. R., Fluker, M. R. & Belcastro, A. N. (1985*b*). The effect of acute exercise on pulsatile release of luteinizing hormone in women runners. *American Journal of Obstetrics and Gynecology*, **153**, 482–5.

Dale, E., Gerlach, D. H. & Wilhite, A. L. (1979). Menstrual dysfunction in distance runners. *Obstetrics and Gynecology*, **54**, 47–53.

Davis, K. & Blake, J. (1956). Social structure and fertility: an analytic framework. *Economic Development and Culture Change*, **4**, 211–35.

Delgado, H., Brineman, E., Lechtig, A., Bongaarts, J., Martorell, R. & Klein, R. E. (1979). Effect of maternal nutritional status and infant supplementation during lactation on postpartum amenorrhea. *American Journal of Obstetrics and Gynecology*, **135**, 303–7.

Delvoye, P., Demaegd, M., Delogne-Desnoek, J. & Robyn, C. (1977). The influence of the frequency of nursing and of previous lactation experience on serum prolactin in lactating mothers. *Journal of Biosocial Science*, **9**, 447–51.

Delvoye, P., Demaegd, M., Uwayitu-Nyampeta & Robyn, C. (1978). Serum prolactin, gonadotropins, and estradiol in menstruating and amenorrheic mothers during two years' lactation. *American Journal of Obstetrics and Gynecology*, **130**, 635–9.

Dietz, W. H., Marino, B., Peacock, N. R. & Bailey, R. C. (1989). Nutritional status of Efe pygmies and Lese horticulturalists. *American Journal of Physical Anthropology*, **78**, 509–18.

Doring, G. K. (1969). The incidence of anovular cycles in women. *Journal of Reproduction and Fertility, Supplement* **6**, 77–81.

Eissa, M. K., Sawers, R. S., Docker, M. F., Lynch, S. S. & Newton, J. R. (1987). Characteristics and incidence of dysfunctional ovulation patterns detected by ultrasound. *Fertility and Sterility*, **47**, 603–12.

Elias, M. F., Teas, J., Johnston, J. & Bora, C. (1986). Nursing practices and lactation amenorrhea. *Journal of Biosocial Science*, **18**, 1–10.

Ellison, P. T. (1981*a*). Threshold hypotheses, developmental age, and menstrual function. *American Journal of Physical Anthropology*, **54**, 337–40.

(1981*b*). Prediction of age at menarche from annual height increments. *American Journal of Physical Anthropology*, **56**, 71–5.

(1982). Skeletal growth, fatness and menarcheal age: a comparison of two hypotheses. *Human Biology*, **54**, 269–81.

(1984). Correlations of basal oestrogens with adrenal androgens and relative weight in normal women. *Annals of Human Biology*, **11**, 327–36.

(1988). Human salivary steroids: methodological considerations and applications in physical anthropology. *Yearbook of Physical Anthropology*, **31**, 115–42.

Ellison, P. T. & Lager, C. (1985). Exercise-induced menstrual disorders. *New England Journal of Medicine*, **313**, 825–6.

(1986). Moderate recreational running is associated with lowered salivary progesterone profiles in women. *American Journal of Obstetrics and Gynecology*, **154**, 1000–3.

Ellison, P. T., Lager, C. & Calfee, J. (1987). Low profiles of salivary progesterone among college undergraduate women. *Journal of Adolescent Health Care*, **8**, 204–7.

46   P. T. Ellison

Ellison, P. T., Lipson, S. F. & Meredith, M. D. (1989a). Salivary testosterone levels in males from the Ituri Forest, Zaire. *American Journal of Human Biology*, 1, 21–4.

Ellison, P. T., O'Rourke, M. T. & Thorne, C. M. (in prep.). Ultradian patterns of salivary progesterone in late adolescent women.

Ellison, P. T., Peacock, N. R. & Lager, C. (1986). Salivary progesterone and luteal function in two low-fertility populations of northeast Zaire. *Human Biology*, 58, 473–83.

(1989b). Ecology and ovarian function among Lese women of the Uturi Forest, Zaire. *American Journal of Physical Anthropology*, 78, 519–26.

Espinos-Compos, J., Rables, C., Gual, C. & Perez-Palacios, G. (1974). Hypothalamic, pituitary, and ovarian function assessment in a patient with anorexia nervosa. *Fertility and Sterility*, 25, 453.

Federation CECOS, Schwartz, D. & Mayaux, M. J. (1982). Female fecundity as a function of age. *New England Journal of Medicine*, 306, 404–6.

Fishman, J. (1980). Fatness, puberty, and ovulation. *New England Journal of Medicine*, 303, 42–3.

Fishman, J., Boyar, R. M. & Hellman, L. (1975). Influence of body weight on estradiol metabolism in young women. *Journal of Clinical Endocrinology and Metabolism*, 41, 989–91.

Fishman, J. & Tulchinsky, D. (1980). Suppression of prolactin secretion in normal young women by 2-hydroxyestrone. *Science*, 210, 73–4.

Frank, O. (1983). Infertility in sub-Saharan Africa: estimates and implications. *Population and Development Review*, 9, 137–44.

Frisch, R. E. (1974). Critical weight at menarche, initiation of the adolescent growth spurt and control of puberty. In *Control of the Onset of Puberty*, ed. M. M. Grumbach, G. D. Grave & F. E. Mayer. New York: John Wiley.

(1976). Critical metabolic mass and the age at menarche. *Annals of Human Biology*, 3, 489–91.

(1978a). Population, food intake and fertility. *Science*, 199, 22–30.

(1978b). Menarche and re-examination of the critical body composition hypothesis. *Science*, 200, 1509–13.

(1980). Pubertal adipose tissue: is it necessary for normal sexual maturation? Evidence from the rat and the human female. *Federation Proceedings*, 39, 2395–400.

(1982). Malnutrition and fertility. *Science*, 215, 1272–3.

(1987). Body fat, menarche, fitness and fertility. *Human Reproduction*, 2, 521–33.

Frisch, R. E., Canick, J. A. & Tulchinsky, D. (1980b). Human fatty marrow aromatizes androgen to estrogen. *Journal of Clinical Endocrinology and Metabolism*, 51, 394–6.

Frisch, R. E. & McArthur, J. W. (1974). Menstrual cycles: fatness as a determinant of minimum weight for height necessary for their maintenance and onset. *Science*, 185, 949–51.

Frisch, R. E. & Revelle, R. (1970). Height and weight at menarche and a hypothesis of critical body weights and adolescent events. *Science*, 169, 397–9.

Frisch, R. E., Revelle, R. & Cook, S. (1971). Height, weight and age at menarche and the 'critical weight' hypothesis. *Science*, 174, 1148–9.

(1973). Components of weight at menarche and the initiation of the adolescent growth spurt in girls: estimated total water, lean body weight and fat. *Human Biology*, **45**, 469–83.

Frisch, R. E., Wyshak, G. & Vincent, L. (1980a). Delayed menarche and amenorrhea of ballet dancers. *New England Journal of Medicine*, **303**, 17–19.

Gardner, J. (1983). Adolescent menstrual characteristics as predictors of gynaecological health. *Annals of Human Biology*, **10**, 31–40.

Gardner, J. & Valadian, I. (1983). Changes over thirty years in an index of gynaecological health. *Annals of Human Biology*, **10**, 41–55.

Goldman, N., Westhoff, C. F. & Paul, L. E. (1987). Variations in natural fertility: the effect of lactation and other determinants. *Population Studies*, **41**, 127–46.

Gomez-Mont, F. (1959). Endocrine changes in chronic human undernutrition. In *Reproductive Physiology and Protein Nutrition*, ed. J. H. Leathem. New Brunswick: Rutgers University.

Graham, R. L., Grimes, D. L. & Campbell, R. D. (1979). Amenorrhea secondary to voluntary weight loss. *Southern Medical Journal*, **72**, 1259.

Green, B. B., Daling, J. R., Weiss, N. S., Liff, J. M. & Koepsell, T. (1986). Exercise as a risk factor for infertility with ovulatory dysfunction. *American Journal of Public Health*, **76**, 1432–6.

Green, B. B., Weiss, N. S. & Daling, J. R. (1988). Risk of ovulatory infertility in relation to body weight. *Fertility and Sterility*, **50**, 721–6.

Greulich, W. W. & Thoms, H. (1939). A study of pelvic type and its relation to body build in white women. *Journal of the American Medical Association*, **112**, 485–93.

Gross, B. A. & Eastman, C. J. (1985). Prolactin and the return of ovulation in breast-feeding women. *Journal of Biosocial Science, Supplement* **9**, 25–42.

Grumbach, M. M., Roth, J. C., Kaplan, S. L. & Kelch, R. P. (1974). Hypothalamic–pituitary regulation of puberty in man: evidence and concepts derived from clinical research. In *Control of the Onset of Puberty*, ed. M. M. Grumbach, G. D. Grave & F. E. Mayer, pp. 115–66. New York: John Wiley.

Haaga, J. G. (1986). Evidence of a reversal of the breastfeeding decline in peninsular Malaysia. *American Journal of Public Health*, **76**, 245–51.

Henry, L. (1961). Some data on natural fertility. *Eugenics Quarterly*, **8**, 81–91.

Hobcraft, J., McDonald, J. W. & Rutstein, S. O. (1985). Demographic determinants of infant and early childhood mortality: a comparative analysis. *Population Studies*, **39**, 363–85.

Hodgson, C. S. (1985a). Using and misusing ethnographic data: lactation as 'nature's contraceptive'. *Medical Anthropology Quarterly*, **16**, 36–40.

(1985b). Response. *Medical Anthropology Quarterly*, **16**, 77–8.

Howie, P. W., McNeilly, A. S., Houston, M. J., Cook, A. & Boyle, H. (1981). Effect of supplementary food on suckling patterns and ovarian activity during lactation. *British Medical Journal*, **283**, 757–9.

(1982a). Fertility after childbirth: infant feeding patterns, basal PRL levels and post-partum ovulation. *Clinical Endocrinology*, **17**, 315–22.

(1982b). Fertility after childbirth: post-partum ovulation and menstruation in bottle and breastfeeding mothers. *Clinical Endocrinology*, **17**, 323–32.

48   P. T. Ellison

Howlett, T. A., Tomlin, S., Ngahfoong, L., Rees, L. H., Bullen, B. A., Skrinar, G. S. & McArthur, J. W. (1984). Release of B endorphin and met-enkephalin during exercise in normal women: response to training. *British Medical Journal*, **288**, 1950–2.

Huang, K. E., Muechler, E. K. & Bonfiglio, T. A. (1984). Follicular phase treatment of luteal phase defect with follicle-stimulating hormone in infertile women. *Obstetrics and Gynecology*, **64**, 32–6.

Hughes, C. L., Fleming, R., Coutts, J. R. T. & Macnaughton, M. C. (1987). Pulsatile hormone secretion late in the luteal phase of normal and infertile women during diurnal hours. *Advances in Experimental Medicine and Biology*, **219**, 629–34.

Jain, A. T., Hsu, C., Freedman, R. & Chang, M. C. (1970). Demographic aspects of lactation and postpartum amenorrhea. *Demography*, **7**, 255–71.

Jelliffe, D. B. & Jelliffe, E. F. P. (1978). *Human Milk in the Modern World*. New York: Oxford University Press.

(1981). 20 million calories are missing. *Lancet*, **1**, 281–2.

(1985). More thoughts on lactation as 'nature's contraceptive'. *Medical Anthropology Quarterly*, **16**, 76–7.

Johnston, F. E., Malina, R. M. & Galbraith, M. A. (1971). Height, weight and age at menarche and the 'critical weight' hypothesis. *Science*, **174**, 1148.

Johnston, F. E., Roche, A. F., Schell, L. M. & Wettenhall, H. N. B. (1975). Critical weight at menarche: critique of a hypothesis. *American Journal of the Diseases of Childhood*, **129**, 19–23.

Keys, A., Brozek, J., Henschel, A. & Mickelsen, O. (1950). *The Biology of Human Starvation*. University of Minnesota: Minneapolis.

Klibanski, A., Beitins, I. Z., Badger, T., Little, R. & McArthur, J. W. (1981). Reproductive function during fasting in men. *Journal of Clinical Endocrinology and Metabolism*, **53**, 258–63.

Knobil, E., Plant, T. M., Wildt, L., Belchetz, P. E. & Marshall, G. (1980). Control of the rhesus monkey menstrual cycle: permissive role of hypothalamic gonadotropin-releasing hormone. *Science*, **207**, 1371–2.

Knodel, J. (1977). Breast-feeding and population growth. *Science*, **198**, 1111–15.

Konner, M. (1985). More thoughts on lactation as 'nature's contraceptive'. *Medical Anthropology Quarterly*, **16**, 73–5.

Konner, M. & Worthman, C. (1980). Nursing frequency, gonadal function, and birth spacing among !Kung hunter–gatherers. *Science*, **207**, 788–91.

Lager, C. & Ellison, P. T. (1987). Effects of moderate weight loss on ovulatory frequency and luteal function in adult women. *American Journal of Physical Anthropology*, **72**, 221–2.

(1990). Effect of moderate weight loss on ovarian function assessed by salivary progesterone measurements. *American Journal of Human Biology*, **2**, 303–12.

Lee, C. S. (1987). Luteal phase defects. *Obstetrics and Gynecology Survey*, **42**, 267–74.

Lee, P. A., Wallin, J. D., Kaplowitz, N., Burkhartsmeier, G. L., Kane, J. P. & Lewis, S. B. (1977). Endocrine and metabolic alterations with food and water deprivation. *American Journal of Clinical Nutrition*, **30**, 1953.

Lev-Ran, A. (1974). Secondary amenorrhea resulting from uncontrolled weight-reducing diets. *Fertility and Sterility*, **25**, 459–62.

Lewis, R., Lauersen, N. & Birnbaum, S. (1973). Malaria associated with pregnancy. *Obstetrics and Gynecology*, **42**, 696.

Liestol, K. E. (1980). Menarcheal age and spontaneous abortion: a causal connection? *American Journal of Epidemiology*, **111**, 753–8.

Lunn, P. G., Austin, S. & Whitehead, R. G. (1984). The effect of improved nutrition on plasma prolactin concentrations and postpartum infertility in lactating Gambian women. *American Journal of Clinical Nutrition*, **39**, 227–35.

Lunn, P. G., Prentice, A. M., Austin, S. & Whitehead, R. G. (1980). Influence of maternal diet on plasma-prolactin levels during lactation. *Lancet*, **1**, 623–5.

McArthur, J. W., O'Loughlin, K. M., Beitins, I. Z., Johnson, L. & Alonso, C. (1976). Endocrine studies during the refeeding of young women with nutritional amenorrhea and infertility. *Mayo Clinic Proceedings*, **51**, 607–15.

MacConnie, S. E., Barkan, A., Lapman, R. M., Schork, M. A., Beitins, I. Z. (1986). Decreased hypothalamic gonadotropin-releasing hormone secretion in male marathon runners. *New England Journal of Medicine*, **315**, 411–17.

McFalls, J. A. Jr. & McFalls, M. H. (1984). *Disease and Fertility*. New York: Academic Press.

McNatty, K. P., Smith, D. M., Makris, A., Osathanondh, R. & Ryan, K. J. (1979). The microenvironment of the human antral follicle: interrelationships among the steroid levels in antral fluid, the population of granulosa cells, and the status of the oocyte *in vivo* and *in vitro*. *Journal of Clinical Endocrinology and Metabolism*, **49**, 851–60.

McNeely, M. J. & Soules, M. R. (1988). The diagnosis of luteal phase deficiency: a critical review. *Fertility and Sterility*, **50**, 1–15.

McNeilly, A. S., Howie, P. W., Houston, M. J., Cook, A. & Boyle, H. (1982). Fertility after childbirth: adequacy of post-partum luteal phases. *Clinical Endocrinology*, **17**, 609–15.

Malina, R. M. (1983). Menarche in athletes: a synthesis and hypothesis. *Annals of Human Biology*, **10**, 1–24.

Maslar, I. A. (1988). The progestational endometrium. *Seminars in Reproductive Endocrinology*, **6**, 115–28.

Masnick, G. S. (1979). The demographic impact of breastfeeding: a critical review. *Human Biology*, **51**, 109–25.

Menken, J., Trussell, J. & Larsen, U. (1986). Age and infertility. *Science*, **233**, 1389–94.

Menken, J., Trussell, J. & Watkins, S. (1981). The nutrition–fertility link: an evaluation of the evidence. *Journal of Interdisciplinary History*, **9**, 425–41.

Metcalf, M. G., Skidmore, D. S., Lowry, G. F. & Mackenzie, J. A. (1983). Incidence of ovulation in the years after menarche. *Journal of Endocrinology*, **97**, 213–19.

Moerman, M. L. (1982). Growth of the birth canal in adolescent girls. *American Journal of Obstetrics and Gynecology*, **143**, 528–32.

Morishima, H., Glasser, B., Niemann, W. & James, L. (1975). Increased uterine activity and fetal deterioration during maternal hyperthermia. *American Journal of Obstetrics and Gynecology*, **121**, 531.

50   P. T. Ellison

Nag, M. (1962). Factors Affecting Human Fertility in Non-industrial Societies: a cross-cultural study. Yale University Publications in Anthropology 66. Yale University Press.
Nillus, S. J. & Wide, L. (1979). Prolonged luteinizing hormone-releasing hormone therapy on follicular maturation, ovulation and corpus luteum function in amenorrheic women with anorexia nervosa. Upsala Journal of Medical Science, 84, 21.
Ojeda, S. R., Advis, J. P. & Andrews, W. W. (1980). Neuroendocrine control of the onset of puberty in the rat. Federation Proceedings, 39, 2365–71.
Ojofeitimi, E. O. (1982). Effect of duration and frequency of breast-feeding on postpartum amenorrhea. Pediatrics, 69, 164–8.
O'Rourke, M. T. (1987). Episodic progesterone release: evidence from salivary measurement. American Journal of Physical Anthropology, 72, 237.
O'Rourke, M. T. & Ellison, P. T. (1990). Salivary measurement of episodic progesterone release. American Journal of Physical Anthropology, 81, 423–8.
Palloni, A. & Millman, S. (1986). Effects of inter-birth intervals and breast-feeding on infant and early childhood mortality. Population Studies, 40, 215–36.
Parker, C. R. & Mahesh, V. B. (1976). Hormonal events surrounding the natural onset of puberty in female rats. Biology of Reproduction, 14, 347–53.
  (1977). Dehydroepiandrosterone (DHA) induced precocious ovulation: correlative changes in blood steroids, gonadotropins and cytosol estradiol receptors of anterior pituitary gland and hypothalamus. Journal of Steroid Biochemistry, 8, 173–7.
Paul, A. A., Mulle, E. M. & Whitehead, R. G. (1979). The quantitative effects of maternal dietary energy intake on pregnancy and lactation in rural Gambian women. Transactions of the Royal Society of Tropical Medicine and Hygiene, 73, 686–92.
Peacock, N. R. (1990). Comparative and cross-cultural approaches to the study of human female reproductive failure. In Primate Life Histories and Evolution, ed. C. J. DeRouseau. New York: Wiley-Liss (in press).
Perez, A., Vela, P., Masnick, G. S. & Potter, R. G. (1972). Ovulation after childbirth: the effect of breastfeeding. American Journal of Obstetrics and Gynecology, 114, 1041–7.
Perez, A., Vela, P., Potter, R. G. & Masnick, G. S. (1971). Timing and sequence of resuming ovulation and menstruation after childbirth. Population Studies, 25, 491–503.
Pintor, C., Genazzani, A. R., Pugoni, R., Carboni, G., Faedda, A., Pisano, E., Orani, T., D'Ambrogio, G. & Corda, R. (1980). Effect of weight loss on adrenal androgen plasma levels in obese prepubertal girls. In Adrenal Androgens, ed. A. R. Genazzani, J. H. Thijssen & P. K. Siiteri. New York: Raven Press.
Pirke, K. M., Broocks, A. & Tuschl, R. J. (1988). Weight reducing diets disturb the menstrual cycle in normal weight healthy young women by impairing episodic LH secretion. Fertility and Sterility, 50, S61.
Pirke, K. M., Schweiger, V., Lemmel, W., Krieg, J. C. & Berger, M. (1985). The influence of dieting on the menstrual cycle of healthy young women. Journal of Clinical Endocrinology and Metabolism, 60, 1174–9.

Polan, M. L., Carding, S. & Loukides, J. (1988). Progesterone modulates interleukin-1 (IL-1) m-RNA production by human pelvic macrophages. *Fertility and Sterility*, **50**, S4.

Polansky, F. F. & Lamb, E. J. (1989). Analysis of three laboratory tests used in the evaluation of male fertility: Bayes' rule applied to the postcoital test, the *in vitro* mucus migration test, and the zona-free hamster egg test. *Fertility and Sterility*, **51**, 215–28.

Popkin, B. M., Bilsborrow, R. E. & Akin, J. S. (1982). Breast-feeding patterns in low-income countries. *Science*, **218**, 1088–93.

Population Reports (1981). *Breastfeeding, Fertility and Family Planning*. Series J, Number 24. Population Information Program. Baltimore: Johns Hopkins University Press.

Potter, J. E. (1987). *Birthspacing and Child Survival: a cautionary note regarding the evidence from the WFS*. Discussion Paper 87–6. Center for Population Studies, Harvard University.

Potter, R. G., New, M. L., Wyon, J. B. & Gordon, J. E. (1965a). A fertility differential in eleven Punjab villages. *Milbank Memorial Fund Quarterly*, **43**, 185–201.

(1965b). Applications of field studies to research on the physiology of human reproduction: lactation and its effects upon birth intervals in eleven Punjab villages, India. In *Public Health and Population Change: current research issues*, ed. M. C. Sheps & J. C. Ridley. Cambridge, Massachusetts: Schenkman.

(1965c). A case study of birth interval dynamics. *Population Studies*, **19**, 81–96.

Prema, K., Nadamuni Naidu, A., Neelakumari, S. & Ramalakshmi, B. A. (1981). Nutrition–fertility interaction in lactating women of low income groups. *British Journal of Nutrition*, **45**, 461–7.

Prentice, A. M., Paul, A. A., Prentice, A., Black, A. E., Cole, T. J. & Whitehead, R. G. (1986). Cross-cultural differences in lactational performance. In *Human Lactation 2. Maternal and Environmental Factors*, ed. M. Hamosh & A. S. Goldman, pp. 13–44. New York: Plenum Press.

Prentice, A. M. & Whitehead, R. G. (1987). The energetics of human reproduction. *Symposium of the Zoological Society of London*, **57**, 275–304.

Prior, J. C. (1985). Luteal phase defects and anovulation: adaptive alterations occurring with conditioning exercise. *Seminars in Reproductive Endrocrinology*, **3**, 27–33.

Ramirez, V. D., Kim, K. & Dluzen, D. (1985). Progesterone action on the LHRH and the nigrostriatal dopamine neuronal systems: *in vitro* and *in vivo* studies. *Recent Progress in Hormone Research*, **41**, 421–72.

Read, G. F., Wilson, D. W., Hughes, I. A. & Griffiths, K. (1984). The use of salivary progesterone assays in the assessment of ovarian function in postmenarcheal girls. *Journal of Endocrinology*, **102**, 265–8.

Reeves, J. (1979). Estimating fatness. *Science*, **204**, 881.

Rich, J. W. (1984). 'Patterns of breast feeding and lactational infertility: a comparison of the Yolngu with the !Kung San and La Leche League.' Thesis, Department of Anthropology, Harvard University.

Rivier, C., Rivier, J. & Vale, W. (1986). Stress-induced inhibition of reproductive functions: role of endogenous corticotropin-releasing factor. *Science*, **231**, 607–9.

52   P. T. Ellison

Rosa, F. W. (1976). Breastfeeding and family planning. *Protein Advisory Group Bulletin*, **5**, 5–7.

Rosenberg, S. M., Luciano, A. A. & Riddick, D. H. (1980). The luteal phase defect: the relative frequency of, and encouraging response to, treatment with vaginal progesterone. *Fertility and Sterility*, **29**, 275.

Sapolsky, R., Rivier, C., Yamamoto, G., Plotsky, P. & Vale, W. (1987). Interleukin-1 stimulates the secretion of hypothalamic corticotropin-releasing factor. *Science*, **238**, 522–4.

Schwartz, B., Cumming, D. C., Riordan, E., Selye, M., Yen, S. S. C. & Rebar, R. W. (1981). Exercise associated amenorrhea: a distinct entity? *American Journal of Obstetrics and Gynecology*, **141**, 662–70.

Scott, E. C. & Johnston, F. E. (1985). Science, nutrition, fat, and policy – tests of the critical fat hypothesis. *Current Anthropology*, **26**, 463–73.

Shangold, M., Freeman, R., Thyssen, B. & Gatz, M. (1979). The relationship between long-distance running, plasma progesterone, and luteal phase length. *Fertility and Sterility*, **31**, 699–702.

Shangold, M. M., Gatz, M. L. & Thyssen, B. (1981). Acute effects of exercise on plasma concentrations of prolactin and testosterone in recreational women runners. *Fertility and Sterility*, **35**, 699–702.

Short, R. V. (1984). Breast feeding. *Scientific American*, **250**, 35–41

Siiteri, P. K., Williams, J. E. & Takaki, N. K. (1976). Steroid abnormalities in endometrial and breast carcinoma: a unifying hypothesis. *Journal of Steroid Biochemistry*, **7**, 897–903.

Sizonenko, P. C. (1978). Endocrinology in preadolescents and adolescents. I. Hormonal changes during normal puberty. *American Journal of the Diseases of Childhood*, **132**, 704–12.

Smart, Y. C., Fraser, I. S., Roberts T. K., Clancy, R. L. & Cripps, A. W. (1982). Fertilization and early pregnancy loss in healthy women attempting conception. *Clinical Reproduction and Fertility*, **1**, 177–84.

Smith, S. R., Chhetri, M. K., Johanson, A. J., Radfar, N. & Migeon, C. J. (1975). The pituitary–gonadal axis in men with protein-calorie malnutrition. *Journal of Clinical Endocrinology and Metabolism*, **41**, 60.

Soules, M. R., Clifton, D. K., Steiner, R. A., Cohen, N. L. & Bremner, W. J. (1988). The corpus luteum: determinants of progesterone secretion in the normal menstrual cycle. *Obstetrics and Gynecology*, **71**, 659–66.

Stanger, J. D. & Yovich, J. L. (1984). Failure of human oocyte release at ovulation. *Fertility and Sterility*, **41**, 827–32.

Stouffer, R. L. (1988). Perspectives on the corpus luteum of the menstrual cycle and early pregnancy. *Seminars in Reproductive Endocrinology*, **6**, 103–13.

Su, T., London, E. D. & Jaffe, J. H. (1988). Steroid binding at σ-receptors suggests a link between endocrine, nervous, and immune systems. *Science*, **240**, 219–21.

Takaki, N. K., Siiteri, P. K., Williams, J. W., Tredway, D. R., Lewis, S. B. & Daane, T. A. (1978). The effect of weight loss on peripheral estrogen synthesis in obese women. *International Journal of Obesity*, **2**, 386.

Tau-Cody, K. R., Campbell, W. F., Dodson, M. G. & Minhas, B. S. (1988). Progesterone markedly enhances *Chlamydia trachomatis* inclusions *in vitro* and results in increased inflammation and productive infection *in vivo*. *Fertility and Sterility*, **50**, S29.

Tho, P. T., Byrd, J. R. & McDonough, P. G. (1979). Etiologies and subsequent reproductive performance of 100 couples and recurrent abortion. *Fertility and Sterility*, **34**, 17.

Thompson, A. M. & Billewicz, W. Z. (1963). Nutritional status, maternal physique and reproductive efficiency. *Proceedings of Nutrition Society*, **22**, 55–60.

Thorne, C. M. (1988). 'Pulsatile progesterone secretion in college age women,' Thesis, Department of Anthropology, Harvard University.

Tietze, C. (1961). The effect of breastfeeding on the rate of conception. *Proceedings of the International Population Conference, New York*, **2**, 129–36.

Toth, A., Senterfit, L. B. & Ledger, W. J. (1983). Secondary amenorrhoea associated with *Chlamydia trachomatis* infection. *British Journal of Venereal Diseases*, **59**, 105–8.

Trussell, J. (1978). Menarche and fatness: reexamination of the critical body composition hypothesis. *Science*, **200**, 1506–9.

(1980). Statistical flaws in evidence for the Frisch hypothesis fatness triggers menarche. *Human Biology*, **52**, 711–20.

Tyson, J. E. (1977). Neuroendocrine control of lactational infertility. *Journal of Biosocial Science, Supplement 4*, 23–40.

Van Ginneken, J. K. (1974). Prolonged breastfeeding as a birth-spacing method. *Studies in Family Planning*, **5**, 201–6.

(1978). The impact of prolonged breastfeeding on birth intervals and on postpartum amenorrhea. In *Nutrition and Human Reproduction*, ed. W. H. Mosley. New York: Plenum Press.

Veldhuis, J. D., Evans, W. S., Demers, L. M., Thorner, M. O., Wakat, D. & Rogol, A. D. (1985). Altered neuroendocrine regulation of gonadotropin secretion in women distance runners. *Journal of Clinical Endocrinology and Metabolism*, **6**, 557–63.

Venturoli, S., Porcu, E., Fabbri, R., Magrini, O., Paradisi, R., Pallotti, G., Gammi, L. & Famigni, C. (1987). Postmenarchal evolution of endocrine pattern and ovarian aspects in adolescents with menstrual irregularities. *Fertility and Sterility*, **48**, 78–85.

Vigersky, R. A., Anderson, A. E., Thompson, R. H. & Loriaux, F. L. (1977). Hypothalamic dysfunction in secondary amenorrhea associated with simple weight loss. *New England Journal of Medicine*, **297**, 1141.

Vihko, R. & Apter, D. (1984). Endocrine characteristics of adolescent menstrual cycles: impact of early menarche. *Steroid Biochemistry*, **20**, 231–6.

Vitzthum, V. J. (1989). Nursing behavior and its relation to duration of postpartum amenorrhea in an Andean community. *Journal of Biosocial Science*, **21**, 145–60.

Wall, S. R. & Cumming, D. C. (1985). Effects of physical activity on reproductive function and development in males. *Seminars in Reproductive Endocrinology*, **3**, 65–80.

Warren, M. (1980). Effects of exercise on pubertal progression and reproductive function in girls. *Journal of Clinical Endocrinology and Metabolism*, **51**, 1150–7.

Warren, M. P., Jewelewicz, R., Dyrenfurth, I., Ans, R., Khalaf, S. & Vande Wiele, R. L. (1975). The significance of weight loss in the evaluation of pituitary response to LH-RH in women with secondary amenorrhea. *Journal of Clinical Endocrinology and Metabolism*, **40**, 601.

54   P. T. Ellison

Wheeler, G. D., Wall, S. R., Belcastro, A. N. & Cumming, D. C. (1984). Reduced serum testosterone and prolactin levels in male distance runners. *Journal of the American Medical Association*, 252, 514–16.
WHO/NRC (USA). (1983). Breastfeeding and fertility regulation: current knowledge and programme policy implications. *Bulletin of the World Health Organization*, 61, 371–82.
Wildt, L., Marshall, G. & Knobil, E. (1980). Experimental induction of puberty in the infantile female rhesus monkey. *Science*, 207, 1373–5.
Woloski, M. R. N. J., Smith, E. M., Meyer, W. J., Fuller, G. M. & Blalock, J. E. (1985). Corticotropin-releasing activity of monokines. *Science*, 230, 1035–7.
Wood, J. W. (1985). More thoughts on lactation as 'nature's contraceptive'. *Medical Anthropology Quarterly*, 16, 75–6.
Wood, J. W., Johnson, P. & Campbell, K. L. (1985). Demographic and endocrinological aspects of low natural fertility in highland New Guinea. *Journal of Biosocial Science*, 17, 57–79.
Wyon, J. B. & Gordon, J. E. (1971). The Khana Study: population problems in the rural Punjab. Cambridge, Massachusetts: Harvard University Press.
Wyshak, G. (1983). Age at menarche and unsuccessful pregnancy outcome. *Annals of Human Biology*, 10, 69–73.
Yen, S. S. C. (1984). Opiates and reproduction: studies in women. In *Opioid Modulation of Endocrine Function*, ed. G. Delitala, M. Motta & M. Serio, pp. 191–209. New York: Raven Press.
Yoshimura, Y. & Wallach, E. E. (1987). Studies of the mechanism(s) of mammalian ovulation. *Fertility and Sterility*, 47, 22–34.
Zeitlin, M. F., Wray, J. D., Stanbury, J. B., Schlossman, N. P. & Meuer, M. J. (1981). *Nutrition–fertility Interactions in Developing Countries: implications for policy and program design*. Washington, DC: Agency for International Development.
Zubiran, S. & Gomez-Mont F. (1953). Endocrine disturbances in chronic human malnutrition. In *Vitamins and Hormones*, ed. P. Harris, U. F. Marian & K. V. Thimann, pp. 9–97. New York: Academic Press.

3 Nutritional status: its measurement and relation to health

C. G. NICHOLAS MASCIE-TAYLOR

Over the past 25 years nutritional status has been increasingly assessed by anthropometry. It is assumed that nutritional problems are manifest as biological effects in individuals, especially growing children. Growth and growth failures are thus an important way to assess these problems (Bogin, 1988). Anthropologists have developed such assessments and these have become an important application of biological anthropology.

Undernutrition is a state of ill-health caused either by an inadequate dietary intake of energy and/or specific nutrients or by excessive losses of protein, etc., in diarrhoeal disease or of iron in haemolytic/haemorrhagic disorders. The four most common types of malnutrition are protein energy malnutrition (PEM), and deficiencies of iron, Vitamin A and iodine (Latham, 1987).

It is thought that PEM affects over 500 million children, most of whom live in the developing countries of Asia, Africa and the Americas, and that 10 million a year die from the direct or indirect effects of PEM. The World Health Organization (WHO) estimate that 350 million women of childbearing age suffer from iron deficiency anaemia. Endemic goitre and cretinism resulting from iodine deficiency affect over 150 million people and in areas where goitre is endemic it is estimated that 1.5 % of the children born have severe mental and physical retardation (cretinism). Thus, 250 000 children are born each year with cretinism solely because of a maternal iodine deficiency. In Asia 6 million people suffer from night blindness caused by vitamin A deficiency (xerophthalamia) and 750 000 people are estimated to die each year from reduced resistance to disease as a result of vitamin A deficiency (Latham, 1984). As these examples show, the links between inadequate nutrition and poor health are only too obvious.

In recent years there has been a decline in the incidence of the more severe forms of PEM, namely kwashiorkor, marasmus and marasmic-kwashiorkor (Latham, 1987), due in some part to improvements in health services and immunization programmes. However, Latham (1984) suspects that the frequency of the less severe forms of PEM has not decreased world-wide. This apparent lack of decline in malnutrition may be explained by many factors, including the continuation of high levels of ignorance and poverty together with political upheavals, wars and natural disasters.

This chapter is not devoted to a reappraisal of these factors; instead, it is concerned with exploring the notions that:

1 there is a clearly defined state of 'malnutrition', in particular, undernutrition, which can be readily measured, and
2 there is a strong relationship between malnutrition, morbidity and mortality.

Measuring nutritional status

To say that a child is malnourished implies that it is beyond or outside the 'normal' nutritional range. But how is nutritional status measured and what determines when a child is classified as malnourished? No sensitive biochemical test exists for measuring nutritional status and although measuring serum protein concentration, especially serum albumin, has been used as an index of protein deficiency, children in developing countries will also be expected to have a deficiency in intake of food as a source of energy which will go undetected by this method.

In the absence of such a test, nutritionists, auxologists, health workers and human biologists have beome increasingly reliant on anthropometric measurements to classify nutritional status. The underlying principle is clear. The most constant and characteristic marker of childhood malnutrition is growth failure. 'Poor growth performance reflects, for the majority of children so designated, deviations from the optimal environmental conditions that support growth' (Haas & Habicht, 1990). The poor environmental conditions that prevail in developing countries lead small children to be exposed to the effects of both an inadequate diet and infectious diseases. Such children might consequently be expected to show a reduction in weight or height from that expected at their age and/ or a reduction in weight for their height.

Use of anthropometric indicators in cross-sectional surveys

During the 1950s and 1960s weight-for-age using the Gomez et al. (1956) classification was the main method used to assess nutritional status both of individuals and of populations. In the 1970s other workers recognized the importance of using both weight and height. In 1971 Seoane & Latham suggested that three different types of malnutrition could be distinguished using knowledge of a child's height and weight. These three types of malnutrition were:

1 Acute, current, short-duration malnutrition where weight-for-age and weight-for-height are low, but height-for-age is normal.
2 Past, chronic malnutrition, where weight-for-age and height-for-age are low, but weight-for-height is normal
3 Acute on chronic long-duration malnutrition, where weight-for-age, height-for-age and weight-for-height are all low.

Nowadays these three types are commonly referred to as wasting (acute), stunting (chronic) and wasting-stunting (acute on chronic) after Waterlow (1972).

Limitations of the NCHS reference curves

The usual way of assessing a child's height and weight is by comparing these anthropometric measurements to a growth reference curve. There are at least six major weight and height (or length) reference curves based on industrialised western countries. These include the Iowa (Jackson & Kelly, 1945), Harvard (Stuart & Stevenson, 1959), Ministry of Health (Ministry of Health, 1959), NCHS (Hamill, 1977), Tanner (Tanner, Whitehouse & Takaishi, 1966) and The Netherlands (Van Vieringen *et al.*, 1971) curves.

In recent years there has been increasing use of the National Center for Health Statistics (NCHS) growth reference curves and the World Health Organization recommended the use of the NCHS reference curves in 1978 (WHO, 1978). These curves are based on American data from the Fels Longitudinal Survey, Cycles 1 and 2 of the Health Examination Survey (1963–70), and the first National Health and Nutrition Examination Survey (HANES 1).

Although extensively used in many developing countries to assess nutritional status, the NCHS curves do have some deficiencies and limitations. For instance, the growth curve for children under 2 years of age was derived from children participating in the Fels Longitudinal Survey who were from white, middle-class backgrounds and it is unlikely that they are representative of all segments of the United States population. In addition, Dibley *et al.* (1987a,b) have found a discontinuity in the data most notably at 24 months of age, when there is an average jump of almost two-thirds of a standard deviation. This discontinuity probably arises from real differences between the two groups of children used to compile the curves as well as from the change in measurement from recumbent length to stature (Trowbridge, Yip & Woteki, 1989). The use of these curves is also complicated because of inconsistency in measurement as an unknown number of children in the age range 24–30 months were measured in the recumbent rather than standing position. In addition, there is an underestimate of height in the age range of 24 to 60 months which arises from inadequate sample size and limitations of the statistical procedures in constructing the growth curves (Yip & Trowbridge, 1989). The Centers for Disease Control (Atlanta, Georgia) are currently correcting these limitations in the growth reference curves (Yip & Trowbridge, 1989). Cole (1989) has suggested the use of the LMS method (Cole, 1988) to overcome problems of skewness and discontinuity.

Reference or standard curves?

The desirability of using a single reference population has also led to a long-running debate partly because of confusion between reference curves and standard curves. A standard implies a target to be aimed at. If little or no difference in the genetic potential for growth in size occurred between children from diverse populations, then the definition of the NCHS curves as standards would be justified. However, if growth differences were dependent to a large extent on gene distributions, then the target or standard for one population would not be the same in each population. The WHO does not appear to accept the NCHS as standards since it refers to them as the NCHS 'reference' (my quotation marks) curves and justifies their use as allowing international comparisons to be made with a universal reference.

Habicht *et al.* (1974) and others contend, however, that the growth of all healthy populations, at least up to 5 years of age, is about the same, thus lending credence to the idea of a universal standard. This view is not shared by some other researchers, for example Goldstein & Tanner (1980). Recently, Eveleth & Tanner (1990) reiterated that a universal standard is inappropriate and suggested that Habicht *et al.* have been misled because of inadequate sampling of populations. Eveleth & Tanner go on to say 'It simply will not do to use an American or a British standard to judge the growth of Japanese or Hong Kong infants or children. Both the size, and the tempo are different.'

The use of the NCHS curves in Africa has also been criticized (Van Loon *et al.*, 1986). Eveleth & Tanner propose that countries should generate their own standards based on well-nourished, healthy individuals drawn from throughout the population and, if continuing past adolescence, should allow for early and late maturers.

A further reason for developing local standards is that the NCHS curves primarily reflect bottle-fed children. Whitehead & Paul (1984) and others have shown that bottle-fed children from developed countries have lower weight gains than infants exclusively breast fed for the first 3 months of life and that the trend reverses for the next 3 months, with faster weight gain for the bottle-fed infants. Consequently, a breast-fed child in a developing country would appear to show above-average growth in the first 2–3 months before its growth rate levels and falls away. In such cases mothers might become worried to an unnecessary degree about their lactational adequacy and introduce supplementary foods, giving up breastfeeding altogether.

In principle, both Eveleth & Tanner and Whitehead & Paul are correct: ideally countries should develop and use their own appropriate norms based on infants whose feeding patterns reflect contemporary and locally appropriate breastfeeding practices. Unfortunately, the practicalities of obtaining growth curves (through longitudinal samples) of truly well-nourished children will defeat the aspirations of many developing countries.

Use of cut-off points

Let us suppose that a nutrition survey is being conducted in a developing country. What cut-off should be used to define and classify a malnourished child? In other words, at what point is a child's height and weight so 'low' as to be classified as outside the 'normal' range?

The simple answer is that any cut-off point is essentially arbitrary. For instance, Gomez *et al.* (1956) used the fiftieth percentile of the Stuart & Stevenson (1959) weight-for-age values as a reference. They then set four categories of nutritional status: normal, defined as greater than 90 % of the median; first degree malnutrition or mild malnutrition between 76 % and 90 % of the median; second degree or moderate malnutrition, 61–75 % of the median; and third degree or severe malnutrition, less than 60 % of the median.

With the NCHS curves, the choice of a cut-off is further complicated because three different statistics – percentile, percentage deviation from the median and the Z score or standard deviation of the median – are commonly used to define a child's relative position and inconsistencies exist in the comparisons of the percentage of reference medians with percentiles and Z scores.

For example, 80 % of the median weight-for-age is equivalent to the tenth percentile for children under 4 months of age but is close to the 5th percentile at 6 months and below the third percentile after 1 year. Thus, compared to a statistical distribution, younger children at 80 % of median are better off than older children at 80 % of median (Haas & Habicht, 1990).

In order to minimize this difficulty, Waterlow et al. (1977) recommended the use of standard deviations (Z scores) from the median of the reference population rather than percentages of the median or percentiles. This procedure is especially useful in studies of growth in which the majority of children fall below the lowest published percentiles for United States children, as is the case with studies conducted in some developing countries. However, even Z scores are flawed conceptually unless a perfect Gaussian distribution is assumed.

Although there is increasing use of Z scores, nutritionists and others have still not reached definitive agreement as to what constitutes a 'low' Z score. In fact, Waterlow's classification is the one which is commonly cited and it has also received the stamp of approval from WHO (1978). The classification divides the distribution into five groups: those with standard deviations (Z) scores for weight-for-height (wt/ht) and height-for-age (ht/age) below -3.0 are severely undernourished; between -3.0 and -2.0, moderately undernourished; between -2.0 and -1.0, mildly undernourished; between -1.0 and ±1.0, adequately nourished; and over ±1.0, overnourished. The cut-off for malnutrition is usually defined by a Z score of -2.0 or below.

The interrelationships for weight-for-height, height-for-age and weight-for-age can be seen in Fig. 3.1. This figure is for 18-month-old boys in the NCHS reference population (WHO, 1983). Using the cut-offs of -2 and ±2 to indicate 'low' and 'high' respectively and M (the median) as 'normal', certain contradictions occur. The distribution of height-for-age is indicated on the horizontal axis. Children to the left of -2 are classified as 'short', those to the right of ±2 as 'tall' and those between -2 and ±2 as 'normal'. The distribution of weight-for-height is shown on the left vertical axis. 'Short', 'normal' and 'tall' children may thus be classified as 'thin', 'normal' or 'obese' in terms of weight-for-height. The two cut-offs of weight-for-age run obliquely across those of height-for-age and weight-for-height. The intersections produce 17 areas than can be described by different combinations of 'low', 'normal' and 'high'. For instance, areas 17, 16 and 14 all refer to children currently underfed,

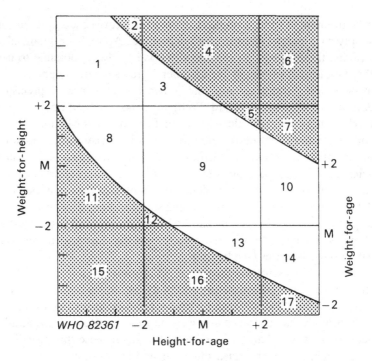

Fig. 3.1. Relationship between Z weight-for-height, height-for-age and weight-for-age. Calculated from the data for 18-month-old boys in the reference population.

whereas area 1 indicates a child currently overfed with a past history of malnutrition.

In the past the most usual indicator of PEM was weight-for-age (Gomez *et al.*, 1956). However, it can readily be seen from Fig. 3.1 that when height is ignored, misinterpretation can result. Take the case of children whose weight-for-age is normal. There are three different interpretations of the status of these children which become apparent if areas 9, 14 and 1 are compared. Children in area 14 are currently underfed as indicated by their low weight-for-height whereas those in area 1 have a past history of malnutrition (low height-for-age) but are currently overfed (high weight-for-height). Only children in area 9 have normal values for all three measurements.

Because of these possible misinterpretations, it is recognized that anthropometric estimates of malnutrition should use indicators that include height. However, Gopalan (1986) states that universal height measurement in India is unlikely in the near future because of the cost of supplying a height measurer in each clinic. Apart from financial constraints, there is another reason why it is not always possible to use all three indicators of weight-for-age, height-for-age and weight-for-height in developing countries and that is because of a lack of precision in determining the age of a child. Frequently, no birth certificates exist and workers have to make do with some form of local calendar. Even in countries where birth certificates are available, registration need not take place until sometime after birth, often leading to concomitant inaccuracies. At best, age is a 'guesstimate'. In these circumstances nutritional status has to be based on weight-for-height and thus it is possible to determine only the present nutritional status of a child without reference to possible past episodes of malnutrition. The use of weight-for-height alone is a constraint which applies to many surveys carried out in developing countries (see below).

Even so, the use of weight-for-height is not without its critics. PEM is characterized by a decrease in both body fat and muscle tissue. A child suffering from PEM will usually have a low weight-for-height, but a tall and normally lean child can also have a low weight-for-height. Similarly, because obesity is characterized by excess fat, an obese child usually has a high weight-for-height but a muscular and large-framed child can also have a high weight-for-height. In other words, excess weight for height does not necessarily imply excess fat and being underweight is not necesarily associated with PEM. In some studies children became more wasted after supplementation because they gained height faster than weight (Rivers, 1988).

One solution to this problem is to measure upper arm muscle area by height (Frisancho & Tracer, personal communication, 1987). In order to compute arm muscle area, mid upper arm circumference (MUAC) and triceps skinfold thickness have to be determined. MUAC has been used in the past as an independent measure of nutritional status (Morley & Woodland, 1979) and more recently by other workers (e.g. Briend, 1987). In addition, the arm circumference : height ratio (the QUAC stick method, Arnhold, 1969) was used as a method of assessing which children were in need of nutritional intervention. The usefulness of upper arm measurements lies in the fact that circumference changes very little with age and between the ages of 1 and 5 years the mean changes less than 1 cm. The arm circumference is usually measured using a Shakir strip which is usually a piece of flexible plastic which does not stretch. If a child's arm circumference at the mid-point of the arm falls below 12.5 cm, this is defined as a measure of severe malnutrition (cf. Briend et al. who used

11.0 cm as their cut-off), between 12.5 and 13.5 cm, moderate malnutrition, and above 13.5 cm good nutritional status. However, inaccuracies of measurement can occur deriving from incorrect determination of the mid point of the arm and of pulling the strip tightly so that the skin wrinkles. Determination of triceps skinfold thickness is notoriously difficult and there are large intra- and inter-observer errors. Consequently, although Frisancho & Tracer's method probably does provide a more complete evaluation of body composition and growth, unless anthropometrists are skilled, measurement errors are very likely. The method is thus not readily usable in developing countries.

Post-famine nutritional survey in Sudan

The practical uses and limitations of the anthropometric approach can be seen from a recent post-famine survey conducted in Sudan. The sub-Sahelian drought in 1983/4 precipitated the most widespread famine recorded in the area. All countries across the sub-Saharan belt were affected and in particular Sudan and Ethiopia. In Sudan the situation was compounded by the large influx of refugees from both Ethiopia and Chad. The famine in Sudan resulted in the largest food relief operation ever undertaken. The US government was the first to recognize the extent and magnitude of the problem and was to become the main food donor.

The urgency in getting food to areas in need meant that there was no time to devise a system of targeting food to the most needy. As a result, blanket food distribution became the norm. Nutritional surveys would have provided a means to select who received food, but only a few were conducted among refugees in east Sudan in 1984. By early 1985 many relief agencies were conducting nutritional surveys, although no effort was made to coordinate activities or use the same procedures to assess nutritional status. This meant that nutritional data could not readily be compared and thus deciding where food should be distributed first was not easy. In view of this, the United States Agency for International Development (US AID) funded a nutritional surveillance programme called the Sudan Emergency Relief Information and Surveillance System (SERISS). The main objectives of the programme were:

1 to assess the nutritional and reported health status of children under the age of 5 at different times and in different areas and administrative units; and

2 to examine the social and economic factors which have a direct and indirect effect on nutritional status.

Background

An Arab proverb observes that Allah laughed when he created Sudan. Another version insists that he wept. Undoubtedly, the original meaning refers to the extremes of physical features found in the area. Nowadays, observers might well feel that the latter version is more correct given the current civil war between north and south Sudan, the severe famine in 1984/5, the continuing precarious economic situation and the recent overthrow of a democratically elected government by an army faction.

The civil war is usually portrayed in the media as a clash between the Arab, Muslim north against the south which is African, Christian and traditional African religions. However, Sudan comprises an ethnically and culturally diverse collection of peoples speaking more than 100 languages and dialects. The 1983 census estimated the population at about 26 million with 60–70 % rural, 10 % nomadic and 20–30 % urban-dwelling. Since then, the combined effects of the famine and war have led to considerable population movements, particularly from south to north and from rural to urban dwelling, so the population of the capital Khartoum has grown by 2–3 million in the past 5 years. Further strain has been created by the influx of a million or so refugees into Sudan from neighbouring countries.

Conducting a survey of the whole of Sudan was not possible given the conflict in the south and so the study was restricted to the six northern regions of Sudan, namely Darfur, Kordofan, Northern, Central, Eastern and Khartoum (Fig. 3.2). Even so, undertaking such a project was not easy. Sudan is an enormous country, the largest in Africa: it spans nearly 18° of latitude and encompasses 2.5 million square kilometres. It is thus nearly four times larger than Texas and 10 times larger than Britain, yet it has only approximately 16 000–18 000 km of roads which comprise an extremely sparse network for the size of the country. Asphalted all-weather roads, excluding paved streets in cities and towns, amount to about 2200 km, of which 1200 km are accounted for by the Khartoum to Port Sudan road which links the capital to the Red Sea. Sudan also has an extensive railway system but it is poorly maintained.

Sampling strategy

Given the size of the country, and even restricting the sample to northern Sudan, it was clear from the outset that it would be logistically impossible

Fig. 3.2. Map of Sudan.

to attempt to visit every village, town and camp. A two-stage cluster design was used which allowed for the relative proportions of the population in nomadic, rural and urban dwellings. In the first of the four rounds of the survey some 15 000 households were to participate from 33 urban clusters, 128 rural clusters and 6 nomadic clusters. Each cluster comprised 90 eligible households in which eligible was defined as any household having a woman who had one or more children under the age of 5, of which at least one was present who could be weighed and measured at the time of the survey. All children under the age of 5 in a household were included in the sample. Because the 1973 census found that each household had 1.7 children under the age of 5, it was anticipated that about 25 000 children would be surveyed. The complete details of the sampling strategy are given elsewhere (Mohamed *et al.*, 1990).

Table 3.1. *Mean and standard deviation of weight-for-height Z scores*

Rounds	N	Mean	SD	% < 2	% < 3
1 May–June 1986	24 041	−1.13	0.99	18.2	2.3
2 Oct.–Nov. 1986	22 431	−1.02	1.03	15.8	2.2
3 Jan.–Feb. 1987	15 538	−0.71	1.00	8.4	0.9
4 May–July 1987	15 799	−0.93	1.01	12.4	1.4
Total	77 809	−0.97	1.02	12.5	1.8

The first round of data collection took place in May and June 1986 and a total of 24 041 children were measured. The second round, from October to November 1986, measured 23 088 different children. At this time it became apparent that little accuracy would be lost if the number of households surveyed per cluster were reduced from 90 to 60. The last two rounds of the survey conducted in January and February 1987 and May to July 1987 used the reduced household numbers and 15 538 and 15 534 children were studied. The total of 78 201 children measured is one of the largest cross-sectional studies of nutritional status ever reported and certainly the largest in a developing country.

Results from SERISS

Each child's height and weight were measured using standard techniques (see Mohamed *et al.*, 1990, for further details). In addition, health over the 2 weeks prior to the survey was recorded and feeding patterns and various household, socioeconomic, and demographic information was obtained (Mohamed *et al.*, 1990).

Because of the problems of inaccuracies in the reported age of children, only the results for NCHS weight-for-height are reported and discussed here. The mean and standard deviation of Z wt/ht scores for each round are presented in Table 3.1 together with the percentage of children falling below −2 and −3 deviations below the mean.

These data provide clear evidence for some degree of acute under-nutrition. In a population of a well-nourished developed country the percentage of children less than −2 standard deviations would be approximately 2.5 % (or 1 in 40) and below −3 about 0.135 % (or 1 in 750). Overall, in the Sudanese under 5s sample, because of a shift of the distribution one standard deviation to the left, approximately 1 in 7 fall below −2 and 1 in 58 below −3 standard deviations.

The overall result masks considerable variation between the four rounds. On average, children gained weight between June/July 1986 and January/March 1987 but they were failing to gain weight relative to height by May/June 1987. This seasonal pattern of results fits in with the harvesting season.

Main predictors of nutritional status

The main predictors of a child's nutritional status were geographical location, age, recent illness and dwelling type (urban, rural or nomadic). A total of 687 clusters were studied in the six regions. The results indicated considerable variation in Z wt/ht both within (the intra-class correlations were very low) and between clusters. However, some general trends were apparent: rural and nomadic children were, on average, more malnourished than urban ones; between birth and 9 months, children were comparatively healthy (Fig. 3.3); after 9 months there was a rapid decline in nutritional status with the worst-affected group aged between 15 and 21 months. Some improvement occurred thereafter.

Feeding patterns were found to be correlated with child age. Even after a multiple regression analysis which had adjusted for the effects of many variables including age of the child, the feeding pattern was still a

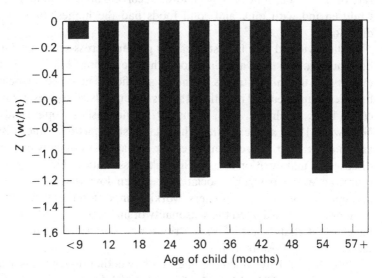

Fig. 3.3. Z weight-for-height by months of age of the child.

significant predictor of weight-for-height. Children exclusively breast-fed had the best nutritional status followed by children receiving breast milk and supplementary food. For weaned children there was a significant decline in nutritional status as the number of meals received per day fell from three to one.

The other main predictor of undernutrition was recent illness. Most children (87.3 %) had suffered from either diarrhoea, measles, respiratory tract infections, vomiting or fever in the 14 days preceding the survey. Children with any one of these conditions were, on average, thinner than children who were unaffected. The differences were particularly marked for children with diarrhoea. Some children suffered from more than one condition (22.7 % suffered from only one condition, 25.1 % two, 17.5 % three and 11.1 % from four or five conditions) and there was a progressive decline in nutritional status as the number of reported conditions increased.

The relationship between malnutrition and diarrhoea has been recently reviewed by Leslie (1987). One of the earliest field studies was conducted by Gordon, Chitkara & Wyon (1963) in the Punjab region of India. They gathered data over a 4-year period on all children born in the study area. The study concluded that, irrespective of age, weaning increased a child's risk of contracting diarrhoea. Their results also showed the importance of breastfeeding in protecting children's health in developing countries since exclusively breast-fed children had the lowest rate of diarrhoea, followed by children receiving breast and other milk. Children fed with milk and solid foods had the highest incidence of diarrhoea.

Leslie reviewed nine field studies that gathered cross-sectional data on both nutritional status and diarrhoeal disease among children. Without exception, the nine studies found a significant negative association between diarrhoea and nutritional status. Four of these studies (Cravioto et al., 1967; Adelman, 1975; Binns, 1976; and Leslie, Gopal Baidya & Nandwari, 1981) also examined the relationship between other disease and malnutrition. Three found either no association between other disease and malnutrition or a much weaker association. However, Binns found an even stronger association between low weight-for-age and pneumonia. Several researchers working in different environmental settings have investigated the seasonality of malnutrition and illness and have found that peaks of malnutrition are associated with peaks of diarrhoea (e.g. Poskitt, 1972).

These studies leave open the question of whether diarrhoea is a cause of malnutrition or whether malnutrition is a risk factor for diarrhoea. In order to show some causal link, seven investigators undertook prospective longitudinal studies to see to what extent infectious disease and particularly diarrhoea is a cause of childhood malnutrition in poor communities in developing countries. A reasonably consistent pattern was found: prior diarrhoea was negatively related to subsequent nutritional status and most studies found that diarrhoea was the only illness to have a significant effect on growth. However, Briend *et al.* (1987) have suggested that the effects of diarrhoea on nutritional status are transitory and that in time catch-up occurs. If their findings are substantiated, then the length of time children are followed up post-diarrhoea is important.

Public health workers in developing countries are usually concerned that malnutrition may increase a child's risk of getting an infectious disease or may increase the duration or severity of an illness. The evidence that the risk of morbidity increases due to malnutrition is not clear (Leslie, 1987). Martorell (1985) believes that 'nutrition interventions have also not been shown to reduce the incidence of infection'. On the other hand there is an association between nutritional status and the severity of infection. Three studies have shown that the length of time spent ill with diarrhoea is consistently greater among the more malnourished children. Sahni & Chandra (1983) also concluded that the most prominent effect of malnutrition on infectious diseases in general was on its duration and severity rather than on its incidence.

The relationship between malnutrition and mortality has also been studied. A study in Bangladesh of children aged 1 to 9 years found that poorer nutritional status significantly increased the risk of dying during the subsequent 18 months (Sommer & Loewenstein, 1975). Although the Bangladesh study found a steadily increasing risk of dying with poorer nutritional status, a study in Narangwal, India (Kielmann & McCord, 1978), found the risk of dying among children was primarily increased in those with severe malnutrition, i.e. those less than 70 % weight-for-age. Chen and his colleagues (Chen, Chowdhury & Huffman, 1980, 1981) also showed that it was the more severely malnourished children according to all anthropometric indices who experienced a markedly greater risk of dying, whereas normally nourished and mildly and moderately malnourished children all experienced smaller but similar risks. These results and others (e.g. Heywood, 1986) point to a variable relationship between nutritional status as measured by anthropometric indices and the risk of dying. Recently, Van Lerberghe (1989) found little or no relationship between anthropometric measures and mortality in a sample from Eastern Zaire.

Other predictors of nutritional status

In addition to the predictors mentioned earlier, some socioeconomic, educational and demographic factors are also associated with the nutritional status of Sudanese children. Their effects can be summarized as follows: on average the child's nutritional status was better when the father was present and had a professional occupation; when the mother had received more than 10 years' education; when there were fewer than average children under the age of 5; when the number of people in the household was large; when per capita grain consumption was in excess of 400 g/person/day; and when the child was a girl. All these results were obtained after adjustment for other variables by multiple regression analysis.

Impact of helminth infections on nutritional status and growth

Although the Sudan post-famine survey collected information on morbidity for the five conditions already mentioned, no information was obtained on whether children were suffering from schistosomiasis, a disease which is endemic in parts of Sudan. At least six studies conducted in Africa have suggested that urinary schistosomiasis may impair child growth, but four others have found no significant cross-sectional relationships between prevalence of *S. haematobium* and children's anthropometric measurements. Some of these apparent differences may arise because of confusion between infection and disease. Although a WHO expert committee used the terms synonymously (WHO, 1981), this view is not shared by other researchers, who draw a distinction between parasitic infection and parasitic disease (Hall, 1985). Thus, an individual with a single paired male and female *Schistosoma haematobium* will be infected but might not show the symptoms of the disease.

The conflicting evidence of the effect of *S. haematobium* on growth and development in these African studies points to the need to measure the intensity of infection, for longitudinal rather than cross-sectional studies, and to select a large number of children and then control for confounding factors such as age, sex, differences in diet and socioeconomic status. Stephenson *et al.* (1985, 1986) overcame most of these potential pitfalls in a Kenyan study and were able to show that the treated group gained significantly more weight than the placebo group over a 6-month period; weight-for-height, weight-for-age, arm circumference and triceps and subscapular skinfold thicknesses were also significantly higher in the treated group.

S. mansoni is known to cause some blood loss into the gut but it is unclear whether this affects nutritional status in individuals who are not severely infected. In studies in St Lucia, heavily infected children had significantly lower serum albumin levels (Cook *et al.*, 1974). Stephenson (1987) suggests that the extent to which *S. mansoni* aggravates or causes human malnutrition is probably underestimated, since evidence from well-controlled animal studies has shown that *S. mansoni* infection causes anaemia, anorexia and weight loss (Mahmoud, 1982).

Besides *Schistosoma* spp., there are other helminths which have important effects on human growth. Intestinal helminth infections are widespread and Peters & Gilles (1977) estimated a quarter of the world's population to be affected. Two types of worm, roundworms and hook-worms, are by far the most important in terms of their impact on the health and normal development of children.

Ascaris infection has been implicated in the cause of malnutrition (e.g. Einhorn & Miller, 1946) and particularly kwashiorkor (Jelliffe, 1953). Several authors have reported a significant association in field situations between *Ascaris* infection and growth. For example, Gupta *et al.* (1977) found that a higher proportion of children treated every 3 months with an anthelmintic gained more than one percentage point in weight-for-age over a year than was the case with children receiving a placebo. Willett, Kilama & Kithamia (1979) observed a 21 % increase in growth in Tanzanian children treated with levamisole compared with a control group receiving a placebo, while Stephenson *et al.* (1980) observed a 33 % increase in weight gain of Kenyan children after removal of *Ascaris* infections. However, Freij *et al.* (1979) found no benefit from periodic deworming in their Ethiopian sample.

The conflicting results from field studies might be partly accounted for by the different anthelmintics used, by sampling and age differences as well as by a failure to control for other factors besides helminth infection, such as socioeconomic variables and other aspects of health status which can confound results (Stephenson, 1987). In addition, none of these studies measured the intensity of infection.

A recent study examined the effects of helminth infections, health and other factors on the growth of 467 pre-school children living in northern Bangladesh. After controlling for the effects of age as well as socio-economic and household variables, children treated with an anthelmintic gained on average 200 g more weight over 6 months and 1.2 cm more in height over 12 months than their untreated counterparts. In addition, treated children experienced significantly fewer measles infections

(Evans, Martin & Mascie-Taylor, 1985) and were less likely to return to supplementary feeding.

The findings from this study based on anthropometric measurements suggest that regular deworming substantially improves the nutritional and health status of pre-school children. Although these measurements provide a good overall view of the impact of intestinal parasitic infection on child growth, more specific tests are required to identify precisely how such infections affect nutritional status.

Apart from the general signs of malnutrition such as reduced rates of growth in both weight and height, Bangladeshi children are particularly prone to deficiencies of iodine, vitamin A and iron, which in severe cases result in goitre, cretinism, zerophthalmia and life-threatening anaemia. *Ascaris* infections have been shown to reduce the absorption of vitamin A by the intestine (Sivakumar & Reddy, 1975; Einagger *et al.*, 1980), thus exacerbating the effect of a dietary inadequacy. This probably explains the association between vitamin A deficiency and ascariasis first reported by Jelliffe & Jung in 1957. Most of the interest in hookworms has naturally centred on their role in causing iron-deficiency anaemia (Lozoff, Warren & Mahmoud, 1975). The worms embed themselves in the duodenal and jejunal mucosa to suck blood. The worms produce an anticoagulant and so when they move to a new site blood continues to ooze from the damaged epithelium. Thus, heavy infections quickly drain the iron stores of the body. In developing countries most of the dietary iron is derived from vegetables rather than meat and is consequently less well absorbed. Thus, the content of most Third World diets cannot compensate for the increased iron loss (Cook, 1976). Apart from their effect on iron status, however, the damage caused by the worm reduces absorption of carbohydrate and protein (Sheehy *et al.*, 1962; Tripathy *et al.*, 1972) and a protein-losing enteropathy may occur (Blackman *et al.*, 1965). Such changes, together with any reduction in appetite, clearly compromise the nutritional status of the host (see review by Solomons & Keusch, 1981).

In experimental animals infected with *Nippostrongylus brasiliensis*, marked changes in intestinal permeability have been observed (Lunn *et al.*, 1986), indicating that the normal function and integrity of the gut has been damaged. In a recent study carried out involving Bangladeshi children infected with *Ascaris*, abnormal intestinal permeability was also observed but had not improved when retested 10 days after antihelminthic treatment (Northrop *et al.*, 1987). However, the rate of intestinal repair is reduced in malnutrition and probably requires a longer time for a significant improvement. These and other animal studies suggest a direct

link between enteric damage by parasitic helminths and growth performance. In the Gambia, Behrens *et al.* (1986) found permeability measurements correlated closely with growth performance. Children with an
abnormal permeability grew much more slowly than those with normal
values. In this study the abnormal permeability was thought to result
from the combination of helminthic infection and disrupted intestinal
mucosa from chronic diarrhoea.

One further possible effect of gastrointestinal parasites is that by
damaging the gut mucosa they may cause plasma leakage and thus loss of
plasma proteins. If severe, this protein-losing enteropathy could lead to
hypo-proteinaemic oedema and the onset of kwashiorkor (Lunn *et al.*,
1986). Although this particular problem is probably more relevant to
hookworm infection (Lunn, 1981), two previous short-term studies
(Martin, Lunn & Wainwright, 1984; Northrop *et al.*, 1987) performed in
Bangladesh, were able to demonstrate a significant improvement in
plasma albumin concentration after treating *Ascaris*. It is clearly important to determine whether long-term protein nutritional status can be
significantly improved by frequent antihelminthic treatment.

The value of anthropometric assessments of childhood nutritional status

At present there is no sensitive biochemical test to assess the degree or
severity of PEM, and so anthropometric measurements will continue to
be used. The measurements usually obtained are weight and height
although some studies use upper arm circumference and skinfold thickness. Skinfold thickness is difficult to measure consistently and is not
recommended for routine use (Owen, 1982).

Anthropometric measurements are ideal for field studies since they are
non-invasive, relatively inexpensive and easy to carry out, although
adequate training is required particularly for accurate determination of
skinfold thickness. The measurement of arm circumference can also be
problematic, especially for children who have recently been immunized
by injection into the left arm (Stephenson, 1987).

The severe forms of PEM (kwashiorkor, marasmus and marasmic-
kwashiorkor) are largely clinically defined, but the mild to moderate
forms have to be assessed in relation to a norm, and this is usually done by
relating a child's height and weight to a reference growth curve. The
reference values most commonly used nowadays are the NCHS reference
values. Increasingly, the NCHS curves have been used, partly because

they are 'approved' by the WHO, but also because they are based on more contemporaneous samples. The use of these international reference values facilitates comparisons between different studies or between groups within the population being studied.

There is confusion between the terms reference and standard. Many researchers refer to the NCHS curves as standards with the implication that they have universal applicability to all populations. Habicht *et al.* (1974) and Martorell (1985) support the view of a universal standard in the pre-school years, but this view has been continually challenged by Tanner and his co-authors (Goldstein & Tanner, 1980; Tanner, 1986; Eveleth & Tanner, 1990), who feel that a universal standard is particularly inappropriate for Japanese and Hong King populations (Leung & Davies, 1989). Instead, Tanner *et al.* recommend the use of local standards.

Some countries, for example India, China and Kenya, have developed their own standards, but producing such growth curves in a developing country can be difficult because large numbers of truly well-nourished children with adequate access to medical facilities are required. Representative groups of children may not be available from all ethnic or geographical segments. Van Loon *et al.* (1986) have recommended basing standards on all children except those showing overt signs of severe malnutrition. This is not to be recommended since children with mild or moderate PEM will be included in the reference group and so the true extent of malnutrition will be underestimated.

Unfortunately, the NCHS curves provide three alternative statistics, percentiles, the percentage of the median and standard deviation of the median (Z scores). All three statistics have advantages and disadvantages but it is increasingly recognized both by academics and by the WHO (Waterlow *et al.*, 1977) that Z scores provide the most accurate and analytically powerful statistic. This view is not shared by Nabarro & Rickleton (1983), who see several disadvantages of using Z scores: they require a knowledge of the reference value for a particular age and they are not easy to explain to people without some statistical background. Nabarro & Rickleton go on to suggest that most health and nutrition workers want to know the size of the difference between, for instance, a child's weight and the reference weight. They do not need to relate that difference to the distribution in the original reference population from North America. Nutritionists and public health workers are becoming more aware of the importance of Z scores and consequently the Z score technique has become increasingly used in field situations. In addition, the United States Center for Disease Control (CDC) has produced a computer program which allows individual measurements to be expressed as Z scores. Many of the difficulties and concerns about the use of the NCHS references have already been mentioned. One further problem is a lack of weight, height and weight-for-height reference values for children over 10 years of age. The NCHS references contain weights and heights for children up to 18 years of age. However, the WHO advises that the references should not be used to compare the nutritional status of groups of children over 10 years of age outside the United States because of considerable differences in the age of onset of puberty in different populations.

In these circumstances body mass index (BMI = weight/height^2), also called Quetelet's index, has been used to compare changes in nutritional status within a community. In Europe mean BMIs of adults fall within the range of 21 to 25 kg/m^2, whereas in developing countries Eveleth & Tanner (1976) found that the mean BMI ranged between 19 and 21 kg/m^2. Although centiles exist (see Cronk & Roche, 1982; Rolland-Cachera *et al.*, 1982), BMI needs to be adjusted for age, particularly in puberty (Van Weiringen, 1972; Rolland-Cachera *et al.*, 1982; Cole, 1986). Norgan (1989) has recently argued that the application of BMI has not been established in developing countries. He goes on to say:

> as BMI is an indicator of size (the amounts of fat and fat free masses) as well as fatness, its relation to fatness and interpretation as a measure of energy stores may vary in different groups. Very low BMI reflects low fat and fat free mass, a state of greater concern than low fat mass alone, and possibly more typical of chronic energy deficiency.

Clearly, more work in developing countries needs to be done to assess the usefulness of BMI in those countries as a measure of nutritional status.

Even so, some view the increasing dependency on anthropometric assessment of nutritional status with some concern. Beaton (1983) argues that 'we have fallen into a trap of assuming that anthropometry is an adequate proxy for "nutrition" rather than an index of one aspect of nutritional status'. He goes on to suggest that:

> In our analyses, we have tended to accept anthropometric status as a proxy for adequacy of food intake in looking at functional outcomes, such as worker productivity or cognitive development . . . achievement of appropriate anthropometric parameters has become the central goal of most nutrition policy and interventions.

He argues there is little or no evidence that body size and composition are intermediary linkages between intake and function and that there is no clear evidence that deviation in anthropometry will necessarily accompany a deviation in function. Energy intake would be expected to serve several functions including maintenance of existing biomass, growth (and pregnancy and lactation), the immune system and physical activity and work. Beaton suggests that anthropometric status may differ without comprising all other functions.

Chandra (1983) has repeatedly suggested that small depressions of anthropometric index are associated with disturbances of the immune system, although it is unclear how these disturbances relate to susceptibility or response to infection. Beaton also cites the data of Rutishauser & Whitehead (1972) from Uganda, where children maintained reasonable growth rates although on apparently low food intakes. A further examination of the community suggested that one way in which the children had achieved this was by a very low level of active play. Chavez & Martinez (1975, 1983) have demonstrated that supplementary feeding increased physical movement among children in a Mexican community.

It can be seen from the foregoing discussions that there are many criticisms, some minor, others more serious, which can be justifiably levelled at the use of the NCHS references in assessing nutritional status. Until new references are constructed, and with little likelihood of a reliable biochemical method becoming available, researchers in a variety of disciplines will have to make do with the NCHS references.

Finally, it is worth remembering that improvements in a child's health can be made inexpensively. Take the case of the deworming project mentioned earlier. In Bangladesh a programme aimed at eradicating intestinal helminth infestations completely from the population would be difficult to implement. Populations of intestinal helminths such as *Ascaris lumbricoides* appear to be remarkably stable and any reduction in the prevalence of infestations which may result from control measures is likely to disappear rapidly once the programme ends (Anderson & May, 1982; Croll *et al.*, 1982). Sanitation is difficult to improve in many parts of Bangladesh since a combination of sandy soils and annual flooding subverts the benefit of latrines. Chemotherapy of the whole population would be expensive and, moreover, unlikely to reduce the community prevalence permanently to zero since infected individuals could continue to immigrate to or pass through the area.

A programme of regular deworming of children is inexpensive (50 cents/child/year) and is relatively easy to operate. The expected improvement in nutritional status as a result of deworming might reduce the need for supplementary feeding, which is more labour intensive, operationally difficult and costs considerably more. In addition, the improvement in intestinal absorption of vitamin A would make a supplementation project more effective in combating xerophthalamia.

References

Adelman, C. (1975). *Health/Nutrition Survey. Kinshasa, Zaire. June 19–July 20, 1974.* Washington, DC: USAID.

Anderson, R. M. & May, R. M. (1982). Population dynamics of human helminth infections: control by chemotherapy. *Nature*, **297**, 557–63.

Arnhold, R. (1969). The QUAC stick: a field measure used by the Quaker service team in Nigeria. *Journal of Tropical Pediatrics*, **16**, 243–7.

Beaton, G. H. (1983). Energy in human nutrition: perspectives and problems. *Nutrition Reviews*, **41**, 325–40.

Behrens, R. H., Lunn, P. G., Northrop, C. A., Hanlon, P. W. & Neale, G. (1986). Factors affecting the integrity of the intestinal mucosa of Gambian children. *American Journal of Clinical Nutrition*, **39**, 112–18.

Binns, C. W. (1976). Food, sickness and death in children of the Highlands of Papua New Guinea. *Journal of Tropical Pediatrics and Environmental Child Health*, **12**, 9–11.

Blackman, V., Marsden, P. D., Banwell, J. & Hall Craggs, M. (1965). Albumin metabolism in hookworm anaemia. *Transactions of the Royal Society of Tropical Medicine and Hygiene*, **59**, 472–82.

Bogin, B. (1988). *Patterns of Human Growth.* Cambridge: Cambridge University Press.

Briend, A. (1987) Detecting children in the community most at risk of dying. *Annual Report of the International Centre for Diarrhoeal Diseases Research, Bangladesh*, pp. 30–1.

Chandra, R. K. (1983). Nutrition, immunity and infection: present knowledge and future direction. *Lancet*, **i**, 688–91.

Chavez, A. & Martinez, C. (1975). Nutrition and development of children from poor rural areas. 5. Nutrition and behavioural development. *Nutrition Reports International*, **11**, 477–89.

(1983). Behavioural measurements of activity in children and their relation to food intake in a poor community. In *Current Topics in Nutrition and Disease. Volume 11, Energy Intake and Activity*, ed. E. Pollitt & P. Amante. New York: Alan R. Liss.

Chen, L. A., Chowdhury, A. K. M. A. & Huffman, S. L. (1980). Anthropometric assessment of energy-protein malnutrition and subsequent risk of mortality among preschool aged children. *American Journal of Clinical Nutrition*, **33**, 1836–45.

(1981). The use of anthropometry for nutritional surveillance in mortality control programs. *American Journal of Clinical Nutrition*, **34**, 2596–9.

Cole, T. J. (1986). Weight/height^p compared to weight/height^2 for assessing adiposity in childhood: influence of age and bone age on p during puberty. *Annals of Human Biology*, **13**, 433–51.

(1988). Fitting smoothed centile curves to reference data (with discussion). *Journal of the Royal Statistical Society, A*, **151**, 385–418.

(1989). Using the LMS method to measure skewness in the NCHS and Dutch National height standards. *Annals of Human Biology*, **16**, 407–19.

78   C. G. N. Mascie-Taylor

Cook, J. C., Bauge, S. T., Warren, V. S. & Jordan, P. (1974). A controlled study of morbidity of *Schistosomiasis mansoni* in St Lucian children, based on quantitative egg excretion. *American Journal of Tropical Medicine and Hygiene*, **26**, 109–17.

Cook, J. C. & Monsen, E. R. (1976). Food iron absorption in human subjects. *American Journal of Clinical Nutrition*, **29**, 859–67.

Cravioto, J., Birch, H. G., De Licardie, E. R. & Rosales, L. (1967). The ecology of infant weight gain in a pre-industrial society. *Acta Paediatrica Scandinavica*, **56**, 71–84.

Croll, N. A., Anderson, R. M., Gyorkos, T. W. & Ghadirian, E. (1982). The population biology and control of *Ascaris lumbricoides* in a rural community in Iran. *Transactions of the Royal Society of Tropical Medicine and Hygiene*, **76**, 187–97.

Cronk, C. E. & Roche, A. F. (1982). Race and sex-specific reference data for triceps and subscapular skinfolds and weight/stature^2. *American Journal of Clinical Nutrition*, **35**, 347–54.

Dibley, M. J., Goldsby, J. B., Staehling, N. W. & Trowbridge, F. L. (1987a). Development of normalized curves for the international growth reference: historical and technical considerations. *American Journal of Clinical Nutrition*, **46**, 736–48.

Dibley, M. J., Staehling, N. W., Niebrug, P. & Trowbridge, F. L. (1987b). Interpretation of Z score anthropometric indicators derived from the international growth reference. *American Journal of Clinical Nutrition*, **46**, 749–62.

Einaggar, B., Graafar, S., Allam, H., Osman, N. & Hussein, L. (1980). Study of the absorption of vitamin A oily preparation among school pupils from the rural Indians. *International Journal of Vitamin and Nutritional Research*, **51**, 3–8.

Einhorn, N. H. & Miller, J. F. (1946). Intestinal helminthiasis: clinical survey of 618 cases of infection with common intestinal helminths in children. *American Journal of Tropical Medicine*, **26**, 497–515.

Evans, J. R., Martin, J. & Mascie-Taylor, C. G. N. (1985). The effect of periodic deworming with Pyrantel pamoate on the growth of pre-school children in Northern Bangladesh. *Save the Children Monographs*, **3**, 1–62.

Eveleth, P. & Tanner, J. M. (1976). *Worldwide Variation in Human Growth*. IBP 8. Cambridge: Cambridge University Press.

(1990). *Worldwide Variation in Human Growth*. (2nd edn). Cambridge: Cambridge University Press.

Freij, L., Meeuwisse, G. W., Berg, N. D., Wall, S. & Gebre-Medhin, M. (1979). Ascariasis and malnutrition. A study in urban Ethiopian children. *American Journal of Clinical Nutrition*, **32**, 1545–53.

Frisancho, A. R. & Tracer, D. P. (1987). Standards of arm muscle by stature for the assessment of nutritional status of children. Personal communication.

Goldstein, H. & Tanner, J. M. (1980). Ecological consideration in the creation and use of child growth standards. *Lancet*, **i**, 582–4.

Gomez, F., Galvan, R. R., Frenk, S., Munoz, J. C., Chavez, R. & Vazquez, J. (1956). Mortality in second and third degree malnutrition. *Journal of Tropical Pediatrics*, **2**, 77–83.

Gopalan, C. (1986). The effect of development programmes on the nutrition of populations. In *Proceedings of the XIIIth International Congress of Nutrition*, ed. T. G. Taylor & N. K. Jenkins. London: John Libbey.

Gordon, J. E., Chitkara, I. D. & Wyon, J. B. (1963). Preventive medicine and epidemiology: weanling diarrhea. *American Journal of Medical Sciences*, **245**, 345–77.

Gupta, M. C., Mithal, S., Arara, K. L. & Tandon, B. N. (1977). Effects of periodic deworming on nutritional status of *Ascaris* infected pre-school children receiving supplementary food. *Lancet*, **iii**, 108–10.

Haas, J. D. & Habicht, J. -P. (1990). Growth and growth charts in the assessment of preschool nutritional status. In *Diet and Disease*, ed. G. A. Harrison & J. C. Waterlow. Cambridge: Cambridge University Press.

Habicht, J.-P., Martorell, R., Yarbrough, C., Malina, R. M. & Klein, R. E. (1974). Height and weight standards for preschool children. How relevant are ethnic differences in growth potential? *Lancet*, **1**(858), 611–14.

Hall, A. (1985). Nutritional aspects of parasitic infection. *Progress in Food and Nutrition Science*, **9**, 227–56.

Hamill, P. V. V. (1977). *NCHS Growth Curves for Children, Birth to 18 Years*. US Department of Health, Education and Welfare, publication number PHS 78–1650. Hyattsville, Maryland: National Center for Health Statistics.

Heywood, P. (1986). Nutritional status as a risk factor for mortality in children in the highlands of Papua New Guinea. In *Proceedings of the XIIIth International Congress of Nutrition*, ed. T. G. Taylor & N. K. Jenkins, pp. 103–6. London: John Libbey.

Jackson, R. I. & Kelly, H. G. (1945). Growth charts for use in pediatrics practice. *Journal of Pediatrics*, **27**, 215–29.

Jelliffe, D. B. (1953). *Ascaris lumbricoides* and malnutrition in tropical children. *Documenta de Medecina Geographica et Tropica*, **5**, 314–20.

Jelliffe, D. B. & Jung, R. C. (1957). Ascariasis in children. *West African Medical Journal*, **6**, 113–22.

Kielmann, A. A. & McCord, C. (1978). Weight-for-age as an index of risk of death in children. *Lancet*, **i**, 1247–50.

Latham, M. C. (1984). *International Nutrition Problems and Policies in World Food Issues* (2nd edn). Ithaca: Cornell University Press.

(1987). Strategies for the control of malnutrition and the influence of the nutritional sciences. In *Food Policy. Integrating supply, distribution and consumption*, ed. J. Price Gittinger, J. Leslie & C. Hoisington. Baltimore: World Bank/Johns Hopkins University Press.

Leslie, J. (1987). Interactions of malnutrition and diarrhea: a review of research. In *Food Policy. Integrating supply, distribution and consumption*, ed. J. Price Gittinger, J. Leslie, & C. Hoisington. Baltimore: World Bank/Johns Hopkins University Press.

Leslie, J., Gopal Baidya, B. & Nandwani, K. (1981). *Prevalence and Correlates of Childhood Malnutrition in the Terai Region of Nepal*. Discussion paper no. 81–35. Population and human resources division. Washington, DC: World Bank.

Leung, S. S. F. & Davies, D. P. (1989). Anthropometric assessment of nutritional status: a need for caution. In *Auxology '88. Perspectives in the Science of Growth and Development*, ed. J. M. Tanner. London: Smith-Gordon.

Lozoff, B., Warren, K. S. & Mahmoud, A. A. F. (1975). Algorithms in the diagnosis and management of exotic diseases. *Journal of Infectious Diseases*, **132**, 606–10.

Lunn, P. G. (1981). Effect of infection on the pathogenesis of protein-energy malnutrition. *Parasitology*, **82**, 41–3.

Lunn, P. G., Northrop, C. A., Behrens, R. H., Martin, J. & Wainwright, M. (1986). Protein losing enteropathy associated with *Nippostrongylus brasiliensis* (Nematoda) infestation of rats and its impact on albumin homeostasis at two levels of dietary protein. *Clinical Science*, **70**, 469–75.

Mahmoud, A. A. F. (1982). Schistosomiasis: clinical features and relevance to hematology. *Seminars in Hematology*, **19**, 132–40.

Martin, J., Lunn, P. G. & Wainwright, M. (1984). Effects of parasitic infection in Bangladeshi children. *Parasitology*, **89**, 64–5.

Martorell, R. (1985). Child growth retardation: a discussion of its causes and its relationship to health. In *Nutritional Adaptation in Man*, ed. K. Baxter & J. C. Waterlow. London: John Libbey.

Ministry of Health (1959). *Standards of Normal Weight in Infancy*. Report on the Public Health of Medical Subjects, Number 99. London: HMSO.

Mohamed, K., Amin, A., Nestel, P. D. & Mascie-Taylor, C. G. N. (1990). Nutritional status of under five year olds in North Sudan: 1. Differences due to geographic location, age, twin status and feeding practices. *Ecology of Food and Nutrition* (in press).

Morley, D. & Woodland, M. (1979). *See How They Grow*. London: Macmillan.

Nabarro, D. & Rickleton, J. (1983). Anthropometric reference figures. In *Annex 2 in Refugee Community Health Care*, ed. S. Simmonds, P. Vaughan, & S. W. Gunn. Oxford: Oxford University Press.

Norgan, N. G. (1989). Body Mass Index and body energy stores in developing countries. *European Journal of Clinical Nutrition, Supplement* **3**.

Northrop, C. A., Lunn, P. G., Wainwright, M. & Evans, J. (1987). Plasma albumin concentrations and intestinal permeability in Bangladeshi children infected with *Ascaris lumbricoides*. *Transactions of the Royal Society of Tropical Medicine and Hygiene*, **81**, 811–15.

Owen, G. M. (1982). Measurement, recording, and assessment of skinfold thickness in childhood and adolescence: report of a small meeting. *American Journal of Clinical Nutrition*, **35**, 629–38.

Peters, W. & Gilles, H. M. (1977). *A Colour Atlas of Tropical Medicine and Parasitology*. London: Wolfe Medical Publications.

Poskitt, E. M. E. (1972). Seasonal variation in infection and malnutrition at a rural paediatric clinic in Uganda. *Transactions of the Royal Society of Tropical Medicine and Hygiene*, **66**, 931–6.

Rivers, J. P. W. (1988). The nutritional biology of famine. In *Famine*, ed. G. A. Harrison. Oxford: Oxford University Press.

Rolland-Cachera, M. F., Sempe, M., Guilloud-Bataille, M., Patois, E., Pequignot-Guggenbuhl, F. & Fautrad, V. (1982). Adiposity indices in children. *American Journal of Clinical Nutrition*, **36**, 178–84.

Rutishauser, I. H. E. & Whitehead, R. G. (1972). Energy intake and expenditure in 1–3 year old Ugandan children living in a rural environment. *British Journal of Nutrition*, **28**, 145–52.

Sahni, S. & Chandra, R. K. (1983). Malnutrition and susceptibility to diarrhea, with special reference to the antiinfective properties of breast milk. In *Diarrhea and Malnutrition: interactions, mechanisms, and interventions*, ed. L. C. Chen, & N. S. Scrimshaw. New York: Plenum Press.

Seoane, N. & Latham, M. C. (1971). Nutritional anthropometry in the identification of malnutrition in childhood. *Journal of Tropical Pediatrics and Environmental Child Health*, **17**, 98–104.

Sheehy, T. W., Meroney, W. H., Cox, R. S. & Solar, J. E. (1962). Hookworm disease and malabsorption. *Gastroenterology*, **42**, 148–56.

Sivakumar, B. & Reddy, V. (1975). Absorption and Vitamin A in children with ascariasis. *Journal of Tropical Medicine and Hygiene*, **78**, 114–15.

Solomons, N. W. & Keusch, G. T. (1981). Nutritional implications of parasitic infections. *Nutrition Reviews*, **39**, 149–61.

Sommer, A. & Loewenstein, M. S. (1975). Nutritional status and mortality: a prospective validation of the QUAC stick. *American Journal of Clinical Nutrition*, **28**, 287–92.

Stephenson, L. S. (1987). *The Impact of Helminth Infections on Human Nutrition*. London: Taylor & Francis.

Stephenson, L. S., Crompton, D. W. T., Latham, M. C., Schulpen, T. W. J., Nesheim, M. C. & Jansen, A. A. J. (1980). Relationship between *Ascaris* infection and growth of malnourished pre-school children in Kenya. *American Journal of Clinical Nutrition*, **33**, 1165–72.

Stephenson, L. S., Latham, M. C., Kurz, K. M., Kinoti, S. N., Odouri, M. L. & Crompton, D. W. T. (1985). Relationships of *Schistosoma haematobium*, hookworm and malarial infections and metrifonate treatment to growth of Kenyan schoolchildren. *American Journal of Tropical Medicine and Hygiene*, **34**, 1109–18.

Stephenson, L. S., Latham, M. C., Kurz, K. M., Miller, D. M. & Kinoti, S. N. (1986). Relationships of *Schistosoma haematobium*, hookworm, and malarial infections and metrifonate treatment to nutritional status of Kenyan Coastal school children: a 16-month follow up. In *Schistosomiasis and Malnutrition*, ed. L. S. Stephenson, pp. 27–65. Cornell International Nutrition Monograph Series No. 16.

Stuart, H. C. & Stevenson, S. S. (1959). Physical growth and development. In *Textbook of Paediatrics*, ed. E. Nelson, Philadelphia: W. B. Saunders.

Tanner, J. M. (1986). Use and abuse of growth standards. In *Human Growth: a comprehensive treatise* (2nd edn), *Vol 3*, ed. F. Falkner & J. M. Tanner, pp. 95–109. New York: Plenum Press.

Tanner, J. M., Whitehouse, R. M. & Takaishi, M. (1966). Standards from birth to maturity for height, weight, height velocity and weight velocity: British children 1965. *Archives of Diseases in Childhood*, **41**, 613–35.

Tripathy, K., Duque, E., Bolanos, O., Lotero, H. & Mayoral, L. G. (1972). Malabsorption syndrome in ascariasis. *American Journal of Clinical Nutrition*, **25**, 1276.

Trowbridge, F. L., Yip, R. & Woteki, C. E. (1989). A revised international growth reference for use in nutritional surveys. In *Abstracts of the XIVth International Congress of Nutrition, Seoul 1989*, Seoul: The Korean Nutrition Society.

Van Lerberghe, W. (1989). Growth, infection and mortality: is growth monitoring an efficient screening instrument? In *Auxology '88. Perspectives in the Science of Growth and Development*, ed. J. M. Tanner. London: Smith-Gordon.

Van Loon, H., Saverys, V., Vuylsteke, J. P., Vlietinck, R. F. & Eeckels, R. (1986). Local versus universal growth standards; the effect of using NCHS as universal reference. *Annals of Human Biology*, **13**, 347–57.

Van Wieringen, J. C. (1972). *Secular Changes of Growth*. Leiden: Netherlands Institute for Preventive Medicine, TNO.

Van Wieringen, J. C., Wafelbakker, F., Verbrugge, H. P. & de Haas, J. H. (1971). *Growth Diagrams 1965 Netherlands*. Netherlands Institute for Preventive Medicine, TNO. Gronigen: Wolters-Noordhoff Publishing.

Waterlow, J. C. (1972). Classification and definition of protein-energy-malnutrition. *British Medical Journal*, **3**, 566–8.

Waterlow, J. C., Buzina, R., Keller, W., Lane, J. M., Nichaman, M. Z. & Tanner, J. M. (1977). The presentation and use of height and weight data for comparing the nutritional status of groups of children under the age of 10 years. *Bulletin of the World Health Organization*, **55**, 489–98.

Whitehead, R. G. & Paul, A. A. (1984). Growth charts and the assessment of infant feeding practices in the western world and in developing countries. *Early Human Development*, **9**, 187–207.

Willett, W. C., Kilama, W. L. & Kithamia, C. M. (1979). *Ascaris* and growth rates: a randomized trial of treatment. *American Journal of Public Health*, **69**, 987–91.

World Health Organization (1978). *A Growth Chart for International Use in Maternal and Child Health Care*. Geneva: World Health Organization.

(1981). *Intestinal Protozoan and Helminthic Infections*. Technical Report Series No. 666. Geneva: World Health Organization.

(1983). *Measuring Change in Nutritional Status*. Geneva: World Health Organization.

Yip, R. & Trowbridge, F. L. (1989). Limitations of the current growth reference. In *Abstracts of the XIVth International Congress of Nutrition, Seoul 1989*. Seoul: The Korean Nutrition Society.

4 Pollution and human growth: lead, noise, polychlorobiphenyl compounds and toxic wastes

LAWRENCE M. SCHELL

> Growth is one of the best indices of child health we have, and a continuous monitoring of the growth and development of children in under-and over-nourished populations is, or should be, a major concern of all public health authorities and governments.
>
> (Tanner, 1966: p. 46)

Environmental contamination from industrialization and modern agriculture is believed to be widespread. In many heavily industrialized countries, the public is partially informed and quite fearful of the biological effects of this contamination (Anderson, 1985). Scientific knowledge of the effects of new substances on health and adaptation has not kept pace with the speed of environmental change. Scientific interest is warranted not only because of public concern, but also because there are questions of biological adaptation to resolve. Environmental contamination is regrettable; however, it presents an opportunity to examine human biological adaptation to a rapidly changing environment. This may help to clarify the relationship between adaptation and community health.

The subject of study, environmental contamination, is an ill-defined phenomenon. Although it can be defined as the effect of pollution, pollution is an equally ambiguous concept. It is recognizable, but has no specific empirical referent. Its definition is, therefore, somewhat arbitrary. Pollution may be defined by source – that is, the product of human vs. non-human activity – but many naturally occurring environmental components (such as volcanic ash or methane) would be excluded by this definition despite their similarity to human products in terms of chemical composition and health effects. Definitions which emphasize the industrial source of pollution are weak also. Many agricultural products themselves are pollutants or are grown with the aid of fertilizers, pesticides or herbicides which contribute to environmental contamination. Nor can the definition be limited to materials, such as chemicals.

Important environmental contamination exists from energy sources such as sound and radiation also. Finally, contamination cannot be delimited by reference to modernity since human contributions to environmental pollution have occurred for millennia, and especially since the development of metallurgy.

For the purposes of this essay, pollution may be defined as materials or energy which are unwanted to some degree, presumably because of interference with human biological well-being. Interference with social, emotional or aesthetic well-being is important also, and is often classified as pollution. However, when such interference is the basis for remediation, it is usually because of a perceived relationship of social, emotional or aesthetic well-being to health. Thus, a direct or indirect threat to biological well-being customarily defines pollutants. This definition complements, on an environmental level, the World Health Organization's definition of health (World Health Organization, 1981).

Pollution defines a new environment for human existence. The environment of many industrial societies is arguably more the product of human activity than any previous environment since the development of agriculture. Even though natural processes contribute greatly to pollutant levels, industrialization is responsible for today's increased human exposure to environmental contamination. In the context of human evolution, this change has been rapid and extensive. There are more new materials, materials at higher concentrations, more commonplace exposure to them, and more exposures at many different ages during the life-span. With the exception of indoor air-pollution from fires (Eisenbud, 1978), and of wine contaminated with lead from underfired pottery (Nriagu, 1983), exposure to pollutants in the past has been to craftspeople primarily, such as metallurgists and ceramicists, and, to the extent that their work occurs in their residence, to their kin as well. Except for occupational environments, contaminant levels were probably low until industrialization. Environmental contamination is now more widespread, and exposure occurs outside the workplace in residential areas. Compared with adults, children are at equal or greater risk for exposure since their play activities may involve close contact with the environment. The foetus is especially at risk since many contaminants pass through the placenta during very sensitive periods of development.

With so many new materials in the physical environment there may be challenges to the survival, reproduction, growth and well-being of humans and other organisms. Challenges to reproduction, and to pre-reproductive survival, may have implications for human evolution. Physiological pathways may be disturbed by many materials that are now commonplace pollutants, and these pathways in turn may affect growth and development. In short, all levels of adaptation may be involved: physiological, ontogenic, genetic and behavioural. Health effects from environmental contaminants that pertain to adaptation have not been comprehensively documented by epidemiologists. However, through a combination of methods from several fields, it is possible to improve both our understanding of human health effects, and of adaptation to contemporary industrial environments simultaneously.

Biological effects of pollution: a multidisciplinary approach

Among the research areas appropriate to this task are auxology and environmental epidemiology, particularly reproductive epidemiology. The latter focuses on threats to spermatozoa, ova, the uterine environment and the developing organism itself. Among these threats are infectious agents, social conditions and environmental contaminants. Environmental epidemiology concerns the distribution of diseases in relation to environmental factors. Auxology is the study of human physical growth and development. Taken together, these fields comprise auxological epidemiology.

Auxological epidemiology is the study of the distribution of patterns of human physical growth in relation to environmental factors, specifically in order to estimate the healthfulness or habitability of environments (Schell, 1984; Tanner, 1986). The premise of auxological epidemiology is that suboptimal growth, growth beneath the genetic potential for growth, reflects decreased environmental habitability. This is based on observations of physical growth's ecosensitivity which date from Boas (1912) and before (see Tanner, 1979), and which have been repeated in numerous studies of various designs (Kaplan, 1954). Morphological plasticity during the growth span is now a recognized feature of *Homo sapiens* (Lasker, 1969).

Auxological epidemiology has been applied most often to assess community health and nutritional status; however, it also can be applied to assess other influences on growth (Schell, 1984, 1986; Tanner, 1986). When nutritional status is good, and genetic differences in growth among populations are minimal, some environmental features may be related to suboptimal growth. Such features may be considered threats to biological well-being.

Auxological and environmental epidemiology can be combined to study the habitability of environments impacted by pollution. This involves combining features of each discipline that are particularly appropriate to study health and adaptation in modern environments. For example, in traditional auxology, great attention is paid to the accurate measurement of growth (Johnston, 1983), and less is paid to measuring accurately the environmental factors which may influence growth. In environmental epidemiology, there is more attention to measuring environmental contaminants and human exposure to them, but little attention is paid to biological changes that are not parts of clinical syndromes.

Epidemiology traditionally focuses on disease rather than on the entire range of biological variation, normal and pathological. This focus is too narrow to assess fully the impact of pollution on health and adaptation. Pollutants may pose biological challenges without producing frank disease, or clinically recognizable syndromes (Colucci et al., 1973). Just as suboptimal nutrition may produce reduced growth without nutritional deficiency disease, pollutant burdens may produce subclinical effects without producing frank disease. Measuring subclinical effects is part of a complete assessment of the biological effects of, and adaptation to, features of the urban environment.

A related problem is that pollution varies considerably in composition. In studies of disease rates, as for example in assessing the impact of toxic waste dumps on the health of nearby communities, it is necessary to specify the disease expected for a given exposure (Hauk, 1982; Landrigan, 1983; Stark, Standfast & Huffaker, 1985). This is usually based on toxicological studies of each of the constituents of pollutants. Since the urban environment includes exposure to many constituents, many outcomes are possible. However, it is impractical to examine subjects for the presence of any and all diseases. On the other hand, specifying one disease would not assess the effects of a complex environment, even though it is necessary for a manageable study. Finally, the epidemiological focus on disease produces difficulties in studying small populations. Because the specific pathologies to be counted are rare events, whether they be diseases (defined syndromes), malformations, stillbirths or neonatal deaths, large populations are needed for adequate statistical power.

Auxological epidemiology meets some of these problems. When the choice of specific health endpoints for community health assessment cannot be guided by theories of pathogenesis or by experience, a general health outcome, such as physical growth, may be used. Growth is based on innumerable physiological pathways and is sensitive to a wide variety of environment factors (Beall, 1982; Bielicki & Welon, 1982; Bielicki, 1986; Eveleth, 1986), Since physical growth status exists as a continuous trait, it may be used to assess the portion of the health pyramid lying below frank pathology. Suboptimal growth may be a more sensitive marker of environmental habitability than counts of specific pathologies. For this same reason it may be the best health outcome for assessing health effects from urban environments. Finally, growth is a feature of every subadult in a population; it is not a rare event as are most epidemiologic measures (malformations, spontaneous abortions, etc.). The statistical power available in studies of growth is large even though specific pollutants may have a more direct and greater effect on the development of very rare pathologies.

There are a variety of other features of growth, unrelated to specific problems in traditional epidemiology, which may make it a suitable health outcome for the assessment of health effects of pollution. A primary advantage is simply the value placed on the health of children. A related methodological characteristic is that growth is assessed without the use of invasive techniques (blood or tissue sampling). Parents and children are familiar with growth assessment through their regular visits to health care providers. An important characteristic of growth assessment is that growth is measured objectively. More confidence can be placed on an objectively assessed health indicator in a community where anxieties about health and safety render untrustworthy self-reports of disease or adverse reproductive events. There are also numerous advantages to growth data when it comes to the time for analysis. Growth data exist on an interval scale with the attendant statistical advantages, and there exists an array of statistical techniques for the analysis of growth data. Finally, the existence of national growth norms allows comparison to a standard population. This is a comparison that is not possible with many of the physiological parameters that may be used to measure health effects from pollutants. For these reasons, physical anthropology can make a contribution to the study of environmental health through the study of human growth in environments of questionable habitability.

The urban–rural comparison

To some extent, the study of urban–rural differences in human growth has begun to answer questions about the effect on growth of environmental contamination. Geographic variation in growth and adult phenotypes have been observed for centuries (Tanner, 1981), and since the mid-nineteenth century anthropologists have recognized an urban–rural difference (see, for example, Topinard, 1890). When differences favoured rural populations, the urban growth depression was sometimes attributed to the enfeebling effects of domestication, and the generally 'unnatural' or ignobling effect of society and culture on human development (see, for example, Prichard, 1813; Topinard, 1890).

Today, however, growth generally is either similar in the two environments, or more favourable in urban children (Eveleth & Tanner, 1976; Eveleth, 1986). However, this does not indicate that the environments are equally favourable for physical growth. There are many social and physical differences between rural and urban environments. Poverty is a potent influence on growth, as the study of slum children shows (Malina et al., 1981; Johnston et al., 1985). Urban–rural differences in growth patterns cannot be attributed solely to differences in the physical or social environments (Schell, 1988). Indeed, with so many factors which can differ, it is difficult in a single study to specify which ones are responsible for the observed biological differences. In some comparisons the factors that strongly influence growth vary little between urban and rural environments (Bielicki & Welon, 1982; Schell, 1988), and growth differences are minimal. However, variation in the factors which distinguish urban and rural environments in several individual comparisons can be exploited when many independent comparisons are brought together. By comparing studies in which the urban–rural environments contrast differently, it may be possible to determine which urban features influence growth. This requires descriptions of both urban and rural environments that are as detailed and accurate as the measure of growth itself.

Case studies

The study of urban–rural differences in growth illustrates two important methodological problems that face auxologists and environmental epidemiologists: the accurate measurement of exposure or dose, and the control of confounding effects. These and other methodological problems are reviewed below as part of a case study of each type of pollutant exposure. With the methodological problems specific to each area, some of the accomplishments of existing studies are described also. The purpose of this review is to expose opportunities for further research.

Air pollution

Nowhere are problems of exposure measurement and confounding more troublesome than in growth studies of communities differing in the degree or type of air pollution. Many growth studies conducted to investigate the health effects of air pollution employ an ecological design.

Two or more areas which actually or presumably differ in air pollution levels are compared. One critical weakness of ecological studies is the presence of other influential exposures which co-vary with the factor of interest. Thus, in studies of air pollution and growth, it is especially important to control for effects of socioeconomic status, a factor which may co-vary with air pollution and is known to exert a strong influence on growth itself (Bielicki & Welon, 1982).

An equally important weakness of the ecological design is the absence of a measure of individual exposure. An assumption of the design is that within each of the groups differing in pollution, all the subjects are equally exposed. At the very least, subjects must not be grossly and systematically misclassified with regard to exposure. Without measuring pollutant burdens in individuals, a large source of variation is ignored. There is no simple remedy. Before area or populational measures of exposure are used to estimate individual exposure, they may need to be corrected for individual characteristics that influence exposure (for example, location of residence within the area, cigarette smoking, occupational exposures). Alternatively, exposure may be measured at the level of the individual by affixing an individual measuring device to each subject. This is an extraordinarily expensive procedure which produces large amounts of data, and its use is limited to prospective studies. Individual body burdens of pollutants may be measured directly in serum or tissue samples, but this method discourages subject cooperation. Alternatively, samples may be taken as part of another survey or for clinical reasons.

Without individual exposure measurement, ecological studies have relied instead on the examination of large numbers of individuals. Large samples can lessen the effect on hypothesis testing of additional variance introduced by random misclassification of subjects' exposures.

Studies of the effects of air pollution on human growth consistently report differences in growth favouring less polluted areas. In a study of births to women in Los Angeles, a significant association was observed between the pollutant level in the area of the city where the mother resided during the pregnancy, and the birth weight of the offspring (Williams, Spence & Tideman, 1977). This study has the advantage of rigorous pollution measurement. In Los Angeles, air pollution levels and the constituents of pollution are measured continuously by environmental monitoring agencies of the government. The effect of cigarette smoking during pregnancy, an important confounder, was controlled in the analysis. A simple measure of socioeconomic status was also included, although this variable may not have been controlled completely.

Nordstrom, Beckman & Nordenson (1978) also observed an effect of air pollution on birthweight. Births to employees at a large smelter and to women from two towns adjacent to the smelter delivered infants significantly smaller in comparison to infants born in the control community and in two other towns distant from the smelter. The effect was greater among more parous gravidae. Although the difference consistently favoured the less-polluted groups, there is little information on the constituents of the air pollution, other environmental parameters, or on other factors that are known to affect birthweight.

Antal *et al.* (1968) compared the growth and morbidity rates among a large sample of children, 4–12 years of age, drawn from two communities differing markedly in air pollution, particularly in lead levels. There was more mobidity of the upper respiratory system in the more polluted city. Weight and height in both cities were advanced relative to national standards but more so in the city with lower air pollution. Thielebeule *et al.* (1980) found slower skeletal maturation and reduced height in a sample of children from a city with high air pollution in comparison to one with cleaner air. A lessening of the initial delay in skeletal maturation and in the degree of height retardation between 1968 and 1978 coincides with a reduction in air pollution in the more polluted city. If no other influences on growth changed concurrently, the improvement in maturation and height may be related to the reduction of air pollution in the previously polluted city.

In a comparison involving more than 10 000 children, 7–12 years of age, Schmidt & Dolgner (1977) found that skeletal maturation was significantly more delayed in city districts with high air pollution than in control areas. They attributed the delay to the effects of reduced ultraviolet radiation exposure in the city with more air pollution. The delay in maturation, however, may be due to other factors such as hypoxia. Children from the air-polluted city also had higher erythrocyte counts and lower levels of haemoglobin per cell. Higher erythrocyte counts and delayed maturation are also observed in some high-altitude populations (Frisancho, 1979). If the children are experiencing mild hypoxia, it may contribute to their delayed maturation. The hypothesis that air pollution resembles other oxygen-limiting stressors agrees with the results from a study of girls, 10–13 years of age, from Silesia (Mikusek, 1976). Girls from the air-polluted town were delayed in skeletal development and in all growth dimensions but chest development. This pattern of selective growth retardation is similar to that seen in some high-altitude populations.

Lead

Lead is known to produce a variety of significant health outcomes in humans exposed to high levels (Damstra, 1977). The effects of chronic, low exposure are less certain, although many studies have reported associations between subacute lead doses and haematological, reproductive, neurological or behavioural outcomes (Nriagu, 1983; Mahaffey, 1985).

Lead is a common pollutant in many environments. In the United States, rural populations have measurable blood lead burdens, although urban populations' levels are substantially higher, and populations residing in the centres of large cities have the highest levels (Mahaffey *et al.*, 1982). Cross-sectional studies indicate that mean lead levels increase during infancy and reach a peak between 2 and 4 years of age (Quah *et al.*, 1982; Hunter, 1986; Kawai, Okamoto & Katagiri, 1987). In the United States, boys have higher levels than girls, and black children have much higher levels than white children (Mahaffey *et al.*, 1982).

Lead is easily absorbed by ingestion and respiration in adults and in children. Children are far more efficient at absorbing ingested lead than adults. Exposure to lead occurs primarily through ingestion rather than respiration. Ingestion of non-food items such as paint or leaded plaster is one source. Another is dust containing lead which precipitates from the atmosphere on to toys, furniture and other surfaces in play areas. From these surfaces it is easily ingested inadvertently.

Even though lead absorption is primarily through ingestion, it may be considered an atmospheric pollutant because the source of most environmental lead is the atmosphere. Among the many components of air pollution, it is one of the more easily and most often measured. Unlike many of the other components of air pollution which leave no evidence of exposure in the body, it is sequestered in the skeleton, circulates in the blood after exposure and can be measured in many internal organs. This provides a major methodological advantage for the study of growth and pollution because it is possible to estimate past and present lead exposure for each individual. Thus, studies of lead burden and growth, for example, are considerably more refined in their design than ecological comparisons of populations residing in areas with grossly different air pollution levels and presumably different lead levels.

The measurement of lead burden has evolved considerably from early measures based on tooth (dentine) level. Using tooth lead partially addressed the problem of assessing chronic past exposure in retrospective and cross-sectional studies of school-age children (Needleman *et al.*,

1974, 1979; Needleman, 1983; Maracek et al., 1983). For all its faults, investigation using tooth lead to gauge exposure has stimulated research on antecedent exposure and outcomes. However, a single, cumulated measure, such as tooth lead, is not sensitive enough to address issues of differential timing of exposure and differential outcomes. Measurement of blood lead addresses this problem. Blood lead reflects current circulating levels, though not current exposure. Circulating levels reflect both current exposure and lead from past exposure which has entered the circulation from the skeletal bank. Blood lead is now the standard means of dose measurement. Thus, Ernhart et al. (1986) and Bellinger et al. (1984, 1986) relate prenatal or cord blood lead to postnatal outcomes, and postnatal lead to postnatal outcomes. However, the temporal specificity of blood lead is also its shortcoming. Although it is measured repeatedly, analyses of lead effects treat each measure as though it were independent. Bellinger et al. (1987), for example, assessed lead at five times in the child's history, and assessed outcomes at 6, 12, 18 and 24 months, allowing for some 20 possible relationships. Relationships with lead assessed at some times and not others can be evidence for a critical period of lead sensitivity (Bellinger et al., 1984, 1986), or of a statistical artifact related to the large number of tests performed (Ernhart et al., 1986). In other words, inconsistencies in the pattern of relationships can arise which are difficult to interpret. Furthermore, when there are numerous measures of lead burden and each is considered separately rather than longitudinally, alpha levels are inflated due to the numerous statistical tests. Also, since the measures of lead burden are intercorrelated, they cannot be considered as independent tests, and the appropriate statistical procedure must be chosen to reduce this error. An alternative approach was taken in the Port Pirie study. McMichael et al. (1986, 1988) measured lead at several times during pregnancy and early childhood, but they integrated the lead levels and used only one cumulative measure of maternal lead burden and child lead burden. This approach reduces the number of tests and the possibility of unpatterned results, but cannot answer questions of critical periods of exposure.

For auxologists, a second problem related to measuring lead burden is the estimation of prenatal lead exposure from maternal blood lead level measured at one or more times during pregnancy. Most studies have either averaged repeated measures of maternal lead or chosen one. This is usually preceded by a preliminary examination of changes in maternal lead by stage of pregnancy. The absence of changes in maternal lead during pregnancy means that any maternal blood lead measurement can stand for the entire pregnancy and it allows the researcher to include any subject with one blood sample drawn before delivery. Since subjects frequently do not attend clinic regularly, the ability to use any subject who attended at least once makes for a much larger sample than if multiple measures had been required. Unfortunately, what little evidence there is about stability of lead levels in infancy (Rabinowitz, Leviton & Needleman, 1984) suggests that levels are highly unstable whether considered as absolute values or as percentiles. Pregnancy may be a period of less physiological stability rather than more, and serial measures may be needed to estimate foetal exposure. Those few studies which have examined changes in blood lead during pregnancy (Barltrop, 1969; Gershanik, Brooks & Little, 1974; Lubin, Caffo & Reece, 1978; Alexander & Delves, 1981) are inconsistent, based on small samples, and cannot take into account the influences of maternal nutrition and anthropometry.

Confounders of a simple relationship of lead burden and child growth and development, including behavioural/cognitive growth, are numerous largely because of the ecosensitivity of growth and development. Confounders are especially important because lead levels tend to be higher in those sub-populations already at risk for growth deficits due to low socioeconomic status, poor nutrition and alcohol or cigarette use.

Few studies have controlled for nutritional status of the child, or in studies of early development, of the mother during pregnancy. Although overt malnutrition is unlikely, iron deficiency anaemia is a likely covariate with lead exposure, and low socioeconomic status. It, too, may affect children's mental abilities (Pollitt, 1987). Calcium is another nutrient that may affect lead uptake, and release from the skeletal bank. Nutritional status generally cannot be ignored in a study of early development. The Cleveland study (Ernhart *et al.*, 1986, 1987) assessed maternal nutrition through a single, 24-hour recall at the second antenatal visit. The Port Pirie study (Baghurst *et al.*, 1985, 1987; McMichael *et al.*, 1986, 1988) included questionnaires on some dietary practices including supplementation for iron, folic acid and calcium as well as dietary calcium (dichotomously coded). In many other studies, nutritional status is not considered even though nutritional needs change during pregnancy and individuals vary considerably in nutrient levels over time. Detailed information on diet and nutritional status on more than one occasion during pregnancy would be a valuable addition to auxological research on lead and growth.

Few studies have examined physical development as an outcome despite its value as a measure of general toxicity on the one hand and child

94   L. M. Schell

well-being on the other. For example, Needleman (1983) observed that height and weight of his high- and low-dentine lead groups were similar, but he did not correct for a statistically significant, 2.5-month age advantage favouring the high-lead group. Reduced physical growth is associated with lead poisoning in children (Damstra, 1977). The effects of low-level chronic exposure, however, are known mostly through anecdotal observation, or from studies without multivariate design or analysis. Habercam *et al.* (1974) observed among a small number of 6–13-year-old black children, highly significant negative associations of blood lead level with height (−0.69), weight (−0.45) and percentiles of height (−0.47) and weight (−0.47), but there was no control for important confounders such as parental size and socioeconomic status. In studies focusing on diet and lead levels, it was also observed that those children with higher levels of lead had attained less height and weight (Mooty, Ferrand & Harris, 1975; Johnson & Tenuta, 1979). In one study of non-organic failure to thrive, Bithoney (1986) noted that subjects' mean lead levels were 22.67 µg/dl vs. 14.3 µg/dl for the controls matched for age, race and income. Some studies of pregnancy outcome have not observed relationships between birthweight and cord blood lead or maternal blood lead at delivery (Rajegowda, Glass & Evans, 1972; Gershanik *et al.*, 1974; Wibberley *et al.*, 1977). However, Fahim, Fahim & Hall (1976) noted that cord and maternal blood lead levels from pre-term births and births with early membrane rupture were significantly higher than levels from normal term births. Likewise, Huel, Boudene & Ibrahim (1981) observed that lead levels measured in hair were higher in mothers of pre-term births than in mothers and infants of normal births. In all of the studies just reviewed, there is little consideration of other factors known to affect physical development, and the sample sizes are very small.

In studies designed to investigate the relationship of lead to postnatal physical growth, it has not been possible to obtain a large sample, extensive knowledge of confounders and measured lead levels. Lauwers and colleagues performed two studies in Hoboken, Belgium. In the first, using mixed longitudinal data, the Preece–Baines model 1 curve was fitted to height and weight means of 1756 girls, 3–18 years of age, from two areas of Hoboken (Lauwers *et al.*, 1984). The areas differed in lead levels in the air (0–2 vs. 2+ µg Pb/m³), but there were no differences between the two groups for any of the curve parameters. Socioeconomic status, as measured by parental occupations, did not differ between the groups, but other influences on growth are difficult to exclude. In a cross-sectional study of 312 children, aged 2.5–16 years of age, blood lead levels were measured, and those with blood lead levels of 40 µg/dl or more were compared to children with levels of 30 μg/dl or less (Lauwers *et al.*, 1986). There was no difference between the lead groups in socioeconomic status as measured by parental occupations. Children in the low-lead group had consistently but only slightly larger values of the 10 anthropometric variables. However, when all dimensions were considered together in a multivariate analysis, the groups differed significantly. A problem with the group definitions is that a blood lead of 20–30 μg/dl is no longer considered low. Lead-associated effects are now believed to appear at levels well below 30 μg/dl (Davis & Svendsgaard, 1987).

An analysis of the NHANES II sample of United States children 0.5–7 years of age (*n* = 2695) found negative relationships between blood lead and height, weight and chest circumference (Schwartz, Angle & Pitcher, 1986). These relationships were observed for children whose lead levels were in the supposedly normal range of 5–35 μg/dl. Although the analysis did control for nutrition, data were lacking on other influences, such as parental size, birth order, family size or socioeconomic status.

In prospective studies of early cognitive/behavioural development, more attention is paid to confounders. For this reason, some reports have included analyses of physical growth, although few studies have focused on growth itself. For the Boston cohort, Bellinger *et al.* (1984) did not observe significant birthweight differences, although this outcome was not analysed through a multivariate analysis. In the Port Pirie study (McMichael *et al.*, 1986), size at birth was analysed through a multivariate analysis, and mothers of pre-term births were found to have significantly higher lead levels than mothers of term births. Among term births, no relationship was found between maternal blood lead and infants' weight, length or head circumference at birth. In the Cleveland cohort, birthweight, length and head circumference were not significantly related to either cord or maternal lead (Ernhart *et al.*, 1986). Since a serious question has been raised regarding the size of the statistical power involved (Needleman, 1987), the absence of an observed relationship in the Cleveland cohort may not be as reliable as in other studies with larger numbers of subjects. In Cincinnati, with a sample not much larger than that of the Cleveland cohort, Dietrich *et al.* (1986) observed that 'prenatal blood Pb was inversely related to birthweight (−157 g per log unit of blood Pb) and to gestational age as well.' Furthermore, Dietrich *et al.* (1987) found that blood Pb was also related to later behavioural/cognitive outcomes. A structural equation analysis indicated a direct effect of lead on the behavioural outcomes and an indirect effect through reduced birthweight and gestational age. (Dietrich *et al.*, 1986, 1987). This suggests that maternal blood lead levels may influence early postnatal growth.

Overall, there appears to be a negative relationship between physical growth and lead burden. There are numerous questions which remain. These concern the critical periods of exposure, the threshold for a lead effect, the permanence of an effect on growth, and the role of nutrition and of other factors in mediating the effect of lead burden on human growth and development. These questions may be addressed more rigorously now because it is possible to measure precisely both lead 'dose' and growth status at the level of the individual.

Chemical exposures: toxic wastes and polychlorobiphenyl compounds

Auxological epidemiology may be able to make an important contribution to knowledge about community health and exposure to chemical wastes. Public health scientists and regulators have depended on epidemiology and toxicology for information to estimate biological effects from chemical waste sites. These fields have provided invaluable information, but there exists some controversy over the ability of traditional study designs to estimate comprehensively biological effects in humans (Bloom, 1981; Lowrance, 1981). Residents near toxic waste sites are usually few in number, and the statistical power of epidemiological methods is compromised considerably in small samples. In addition, residents of toxic waste site communities are often emotionally strained and their reports of symptoms and health effects may be biased. The only reliable information may be from existing tumour registries or death records, both of which may reflect conditions many years earlier.

Some limitations of toxicology are reached quickly in the study of chemical waste site effects. The toxicological approach to determining biological effects depends upon knowing which chemicals are present at the site, and estimating the effect from each chemical. Learning which chemicals are present requires very costly tests, and some chemicals, such as TCDD (2,3,7,8-tetrachlorodibenzo-p-dioxin), may be present as trace contaminants measured in parts per billion, or in even smaller fractions. Laboratory determinations must have been made of dose levels for lethal and sublethal effects for all the chemicals present. The predicted dose–response relationship may not include synergistic effects of the chemicals mixed at the site. Even if common mixes could be duplicated in the laboratory, it remains necessary to use non-human models to estimate effects in humans. Finally, in most cases of human residential exposure, the exposure itself is ill-defined, and predictive models are difficult to apply to individuals or to neighbourhoods.

Auxological epidemiology can address some of these problems. It investigates the stress as it exists in the environment, rather than as approximated in the laboratory. The outcome variable, growth, is objectively measured and not affected by subject biases. Large numbers of subjects, relative to the community's size, are obtained by including all births and all children as data points. Physical growth is a general health outcome, being the product of many physiological pathways which may be altered by chemicals present at toxic waste sites. Finally, the growth of children is a measure of their physical well-being, and is a measure of health that is valued by the community itself. An important limitation is that this approach cannot specify causal relationships. However, evidence of an association between growth and toxic chemical exposure is the starting point for other types of studies designed to elucidate causal pathways.

There are few studies of health effects from toxic waste or chemical exposure that include a consideration of the growth of children. The most informative organization of the available information would be by type of exposure – that is, by chemical – or by exposure period in the life-span. However, the available information on human growth and toxics cannot be organized by either scheme as can toxicological studies of controlled exposures of laboratory animals. Dietary and residential exposures, the ones which are most likely to involve both children and pregnant women, are usually to mixtures of chemicals. Occupational exposures offer more specificity. The exposure is usually limited to adults, including women of childbearing age, and the variety of chemicals in a particular workplace may be smaller than in residential exposures to uncontrolled chemical waste emanating from dumpsites. The available information is categorized as residential exposure to a heterogeneous mix of chemicals, or as occupational exposure.

Physical growth has been investigated among residents of Love Canal, New York, where a large variety of waste chemicals emanating from a dumpsite presented an ill-defined exposure to residents. The canal itself, shallow and 3000 m long, received chemical wastes between the early 1940s and 1953. Approximately 19 000 metric tonnes of 248 different waste chemicals were deposited. After dumping stopped and the canal was covered and graded, a school was built on the site.

Exposure estimates for individual residents are problematic. Use of the school area for play, proximity of residence to the canal, location of residence *vis-à-vis* preferential routes of chemical migration, duration of residence, and residence during certain types of dumping, all may influence exposure. In addition, past exposure could not be indicated by measuring body burden. For example, polychlorinated biphenyls, which are normally deposited in adipose tissue, were not present in the waste, and could not be used as a marker of past chronic exposure to the wastes generally.

Vianna (1980) presented some preliminary findings on reproductive performance of women at Love Canal. He noted that there was an excess of spontaneous abortions among Love Canal residents thought to have been at greater risk for exposure to the chemical leachate. Residents of 'wet area' homes, built on swales which may have acted as preferential routes for the migration of chemicals, had more adverse pregnancy outcomes when compared to a standard population and compared with Love Canal residents thought to have had less exposure in non-water area homes built away from the swales. Of the 460 live births in the study, 10.6 % were low birthweight infants, but the rate in the wet area homes was twice that of the dry area homes (13 of 83 births versus 11 of 144 births). Independently, Goldman, Paigen and colleagues also examined the frequency of low birthweight, and of children's health problems at Love Canal (Goldman et al., 1985; Paigen et al. 1985). Using multivariate techniques, the analysis took into account such important influences on foetal growth as socioeconomic and demographic characteristics, and other factors such as cigarette smoking and alcohol use. For homeowners, living near the canal was strongly associated with babies of low birthweight (Goldman et al., 1985). Of 57 babies born to homeowner households in 'wet' (swale) homes, 10 were of low birthweight. This is twice the rate found among the control group. Dry households did not have a higher rate compared with control households. There were no differences in gestation ages, but a significant increase in the rate of parent-reported birth defects.

A serious problem for any study of exposure to toxic waste is the completeness of the sample, particularly the inclusion of the most heavily exposed individuals. Many of the most exposed residents relocate away from the source of pollution, and are difficult to include in a community-based study. At Love Canal, the families living closest to the dumpsite were evacuated, and many left the area. The New York State Department of Health sought every adult female who had resided in the Love Canal neighbourhood between 1940 and 1978, some 1295 subjects, for a study of low birthweight (Wianna & Polan, 1984). Ninety-three per cent were included. There was a significantly higher rate of low birthweight among infants born to women who had resided in swale areas (wet homes), and presumably had greater exposure to leachate from the canal dump. The rate of low birthweight among swale households was nearly twice the rate among homes abutting the canal, and the rate was significantly higher than the rate for residents of the rest of the canal. Analysis of the timing of the low birthweights showed a marked excess between 1946 and 1958 during which 8 of 18 live births had low birthweight. During the period of active waste dumping, the percentage of low-weight births in the swale area was significantly higher than in the rest of New York state, excluding New York City. It was also higher than in the rest of the canal neighbourhood which, in turn, did not differ in the percentage of low-weight births from the rest of upstate New York during this same time.

Postnatal physical growth may have been affected also. Paigen *et al.* (1987) found that children who were both born in the Love Canal area and had spent 75 % or more of their life there ($n = 170$) were significantly shorter than controls ($n = 610$). These differences in size at birth and in postnatal size were not attributable to confounding factors such as socioeconomic status, cigarette use or race. These had been controlled in the study design, or in the analysis with multivariate statistical methods.

In all studies of Love Canal residents, there remains some uncertainty about exposure. The findings reviewed here were obtained without detailed information about exposure to specific chemicals. Exposure was not estimated by measuring individuals' body burdens, but with two crude measures of exposure, distance from the canal and wet versus dry homes.

Physical growth and development has also been examined among Japanese who had consumed contaminated rice oil in 1968. During production, the rice oil had become tainted with a mixture of polychlorinated biphenyls (PCBs), polychlorinated dibenzofurans and polychlorinated quaterphenyls (Masuda & Yoshimura, 1984). Approximately 1800 Japanese consumed some of the oil, and many developed the disease Yusho. In 1979 a similar outbreak occurred in Taiwan, also traced to rice oil which contained a variety of polychlorobiphenyl compounds. Yusho is characterized by several non-specific symptoms (fatigue, weight loss, anorexia) and a variety of more specific ones, many of which involve dermatological signs such as extra pigmentation and chloracne (Okumura, 1984). Symptomology is difficult to relate to exposure since the exact dose for each Yusho victim is not known. Exposure varied with the amount of oil consumed and the varieties of PCBs and other toxic materials it contained. Yusho was originally believed to be caused by PCBs, but with the discovery of PCB contaminants, the polychlorinated dibenzofurans and quaterphenyls, blame for the symptoms has shifted to the more toxic dibenzofurans (Masuda & Yoshimura, 1984). Thus, Yusho should not be considered a PCB poisoning episode, and the symptoms may differ from PCB poisonings through an occupational or dietary exposure.

Effects on growth can begin early in life. Polychlorobiphenyls are lipophilic, and cross the placenta easily (Kuwabara et al., 1978; Jacobson et al., 1984). Correlations between maternal and foetal serum PCB levels range from 0.42 to 0.71 (Jacobson et al., 1984; Roncevic et al., 1987). PCB levels in breast milk are highly correlated with maternal serum levels (Kuwabara et al., 1978; Hara, 1985). Mothers with a PCB burden who breastfed their children transferred PCBs to the child (Yoshimura, 1974; Hara, 1985). In one study of children of mothers with an occupational exposure to PCBs, those children breastfed for 3 months or more had nearly 10 times the PCB levels of children who were not breastfed, and nearly twice that of their mother's serum PCB level (Yakushiji et al., 1984).

Infants born to Yusho mothers may have a characteristic syndrome (Miller, 1985). A very common feature of this syndrome is small size at birth (Taki, Hisanaga & Amagase, 1969; Funatsu et al., 1971; Yamashita & Hayashi, 1985). Yusho children exposed postnatally grew at lower rates than unexposed children, although in some studies the difference was statistically significant only among males (Yamaguchi, Yoshimura & Kuratsune, 1971; Yoshimura, 1971; Fujisawa & Fujiwara, 1972; Yoshimura & Ikeda, 1978). Some Yusho-related retardation may be permanent. Yoshimura & Ikeda (1978) found that 4 years after the Yusho exposure, yearly growth increments were similar, but the affected children had not accelerated in growth to catch up with their unexposed peers.

Some uncertainty remains regarding whether PCBs are responsible for the reduced growth among Yusho children, since dibenzofurans and quaterphenyls may have been ingested also. The possibility that exposure to PCBs alone may retard growth has been investigated in studies of women occupationally exposed to PCBs. In such studies, it is presumed that the exposure is restricted to just one chemical or, at very least, to fewer chemicals than in accidental exposures such as at Love Canal or that which caused Yusho in Japan and Taiwan. Taylor et al. (1984) compared births of two groups of women: those having physical contact with PCBs at work for at least 1 year prior to the birth, and those who had worked in the same plant but where PCBs were not directly used. Exposure was not estimated through analysis of tissues for PCB levels, and the presence or absence of direct contact was the sole measure of exposure. Direct contact can be an important differentiator of exposure because PCBs can be absorbed dermally. The women with direct contact had babies significantly lighter in weight (by 153 g) and with gestations 6.6 days shorter than babies of women without physical contact. After adjustment for maternal age, parity and the sex of the infant, the negative association remained. A follow-up study which included the original sample collected additional information from interviews to control for several influences upon prenatal growth such as maternal height, weight and cigarette smoking. This investigation detected small but statistically significant effects of PCB exposure on gestational age and birthweight, with the latter effect only partly mediated through gestational age effects (Taylor, Stelma & Lawrence, 1989).

Control over confounding factors is difficult in studies relying entirely on records. Information about such confounding factors as cigarette smoking are often absent. Accurate estimates of pollutant body burden (exposure) are virtually impossible to obtain also. Retrospective study designs which do not rely solely on records may be more useful. In a study of 912 North Carolina infants, mother's serum and breast milk were analysed for PCB levels (Rogan *et al.*, 1986). Birthweight, head circumference and some scores on the Brazelton Neonatal Behavioural Assessment Scale were unrelated to the mother's cumulated body burden of PCBs. In the multiple regression analysis of birthweight, only cigarette smoking, sex and mother's weight were significantly related to birthweight. Maternal age, education, occupation, race, gravidity and a measure of alcohol use were not included in the regression model.

Women and children may be exposed to PCBs in their diet. PCBs have contaminated food on several occasions in the United States (Waldbott, 1978). Regulations are now in place which set the permissible levels of PCBs in commercial food products in the United States making chronic exposure through the consumption of highly contaminated foods less likely now. However, a remaining exposure route is through the consumption of such unregulated foods as game fish. One study of women who varied in their consumption of fish from Lake Michigan related PCB exposure to newborn size, gestational age and maturity (Fein *et al.*, 1984). Three measures, or models, of exposure were employed. In the first, PCBs were directly measured in serum from the umbilical cord at the child's birth. PCB exposure was also measured as the maternal intake of PCBs through the consumption of fish from Lake Michigan, weighing the PCB contributions of each species by the average contaminant level of that species in Lake Michigan. With this technique, the second and third exposure measures were constructed: overall maternal ingestion and ingestion only during pregnancy. Overall fish consumption showed a consistent dose–response relationship with birthweight, head circumference, gestational age and neuromuscular activity. Infants with detectable levels of PCBs in their cord serum had lower birthweights, smaller head circumferences and shorter gestational ages. However, ingestion only during pregnancy was not related to measures of infant size. This suggests that the PCB burden accumulated before pregnancy and measured in serum is more closely related to prenatal growth than the added exposure from intake of PCBs just during the pregnancy.

With several important exceptions, studies of humans exposed to waste chemicals through their diets, residences or occupations have found associations with retarded physical prenatal and postnatal growth. However, the exceptions to this generalization cannot be dismissed. In general, there has not been sufficient replication of results to reach a conclusion about causal factors. While studies of postnatal growth retardation at Love Canal and among Yusho children are consistent, they are few in number. The inconsistency among studies of prenatal growth can be attributed to many factors. Studies differ in the level of the subjects' exposure, the methods used to estimate exposure and the amount of control for confounding factors.

In most observational studies (that is, without randomization of exposures), some doubt remains concerning the control of confounding factors (Campbell & Stanley, 1963). Replication of studies with populations differing in the complex of confounders is often the only solution. Variation in actual exposure may not be a drawback since the results from different studies may be used to approximate a dose–response curve. These curves are helpful for epidemiologists to establish causal relationships (McMahon & Pugh, 1970).

Of the three factors contributing to inconsistency among studies, uncertainty about actual exposure is the most difficult one to resolve. Since studies are usually conducted after the exposure has taken place, retrospective exposure assessment is difficult. A social science solution is to collect information about past behaviours associated with exposure risk, and, using an *a priori* model of exposure, to estimate chemical burdens for individuals (Fein *et al.*, 1984), or for groups (Vianna, 1980; Taylor *et al.*, 1984; Vianna & Polan, 1984; Goldman *et al.*, 1985; Paigen *et al.*, 1987). A biological marker approach to estimate past exposure is to measure chemicals which bioaccumulate, such as lead and PCBs (Fein *et al.*, 1984; Rogan, Gladen & Wilcox, 1985). This application is limited to measuring chemicals which bioaccumulate, or using them as markers for those which do not. This assumes that there is a consistent ratio of the two types of chemicals in each human exposure. On the other hand, the social science, or risk behaviour modelling approach, assumes that their model of exposure accurately identifies the behaviours creating a risk of exposure, that people can remember those behaviours, and that they recall them without bias.

Noise

Noise pollution differs from most other pollutants in that noise is energy rather than a material. Furthermore, since noise is usually defined as unwanted sound, it is energy which much be interpreted. Nevertheless, in most studies of its biological effects, the distinction is moot.

Effects of noise can be divided into two types: auditory and non-auditory. Auditory effects include permanent and temporary shifts of the threshold for hearing. Noise-induced shifts generally begin at the higher frequencies. Temporary and permanent threshold shifts can be caused by acute or chronic exposure to loud noise. These shifts can occur at any age if noise exposure is great, and their effect is not only upon the ear, but may affect classroom communication and social communication generally. Non-auditory effects are a heterogeneous group including psychological disturbances and a variety of physiological alterations. Non-auditory effects may be produced by chronic exposure to sound at levels too low to effect threshold shifts (Cantrel, 1974). Indeed, intact hearing is a prerequisite for responsivity at sound energy levels below those capable of producing direct physical damage. Effects on psychological well-being has been documented by comparing rates of admission to mental hospitals (Kryter, 1987) and rates of the use of medications which alleviate anxiety and stress (Knipschild & Oudschoorn, 1977; Watkins, Tarnopolsky & Jenkins, 1981). Physiological effects of noise exposure are related to the stress response. Numerous studies have shown that acute noise is a physiological stressor stimulating the autonomic nervous system and the adrenal cortex, and causing the complex chain reaction of effects on the cardiovascular, gastrointestinal, endocrine and reproductive systems (for review see Welch & Welch, 1970; Schell & Lieberman, 1981; Westman & Walters, 1981; Kryter, 1985).

Laboratory evidence of sustained physiological alterations following exposure to chronic noise is more equivocal. The strongest evidence for chronic, health-related effects comes from studies of workers chronically exposed to noise (for example, Cohen, 1968, 1973), and of residents near airports (Karogodina *et al.*, 1969; Knipschild, 1977*a,b*; Cohen *et al.*,

1981). Because noise is a physiological stressor, and stressors produce a common constellation of responses, studies of chronic noise exposure may inform us of the effects of sustained stress generally, as well as of the non-auditory effects of noise itself.

The relationship between human physical growth and high noise levels has been examined in a study of occupational exposure (Hartikainen-Sorri *et al.*, 1988) and through studies of residents near airports in Amsterdam, Paris, Japan and the United States. All studies have cross-sectional or entirely retrospective research designs with the usual problem of exposure misclassification. The alternative, individual dosimetry, is impractical for the study of large numbers of children or pregnancies and cannot be used retrospectively. A prospective design also is impractical for the study of chronic stress as it would involve years of follow-up and measurement. The strategy of airport noise studies is to compare extremes of past exposure and so minimize exposure misclassification.

The extent of information regarding actual noise levels experienced by the subjects varies from study to study. In some, noise exposure is estimated from recalled self-assessments (Hartikainen-Sorri *et al.*, 1988), and in others from government agencies which routinely monitor noise levels and whose measurements bracket the period when the children were born (Schell, 1981, 1984; Schell & Hodges, 1987). In still other studies, noise levels may not have been measured, and high noise exposures are presumed, based on the proximity of the residences to the airport runways.

An additional problem is the presence of other influences on growth which co-vary with noise exposure. Particularly important among these is socioeconomic status. Preliminary analyses do not take these into account fully (Coblentz & Martel, 1986), and unless the control group is matched closely, adjustment for confounding factors is made through multivariate statistical analyses.

In observational studies of noise and growth, replication adds some certainty to the conclusions. In each study, there are differences in the social factors influencing growth, in the degree of control over confounders, in exposure assessment and in the noise levels themselves. Studies of airport noise and prenatal growth are consistent: smaller birthweights are associated with greater noise exposure (Ando & Hattori, 1973, 1977; Knipschild, Meijer & Salle, 1981; Schell, 1984; Coblentz & Martel, 1986; Schell & Hodges, 1987). There is a shift in mean birthweight (Ando & Hattori, 1973, 1977; Schell, 1984; Coblentz & Martel, 1986; Schell & Hodges, 1987), or an increase in the frequencies of lower birthweights (Knipschild *et al.*, 1981). The exception is a study of occupational exposure (self-assessed) which did not observe an increased risk of prematurity, or of birthweights less than 2500 g (Hartikainen-Sorri *et al.*, 1988).

Studies of postnatal growth are few, and the results from these are not as consistent. Takahashi & Kyo (1968) studied children of 1–16 years of age who were exposed to airport noise in Japan. Primary school-aged boys were significantly shorter than their peers, but the difference was not statistically significant among the girls. When there were differences among older children and youths, they favoured the less-exposed children also. In the author's study of 5- to 13-year-old children living near an international airport in the United States, children from the noise-exposed community were slightly leaner than children in the control group (Schell & Norelli, 1983). Height, shoulder, hip and facial breadths were similar between the exposed and non-exposed children. Analysis of the change in size between birth and the age of postnatal growth assessment suggested that some catch-up growth may have taken place among the most noise-exposed children (Schell & Hodges, 1985).

Questions remain regarding the effects of noise stress on human reproduction, physical growth and health. In the design of additional non-laboratory studies, there are two interrelated concerns: the size of the samples and the intensities of the noise exposures. Large samples chronically exposed to moderate levels of noise may not be adequate. Results from studies where there is a range of exposures suggests that moderate and low levels are not associated with effects on growth (Knipschild *et al.*, 1981; Schell & Norelli, 1983; Schell, 1984; Schell & Hodges, 1985). Clarification of effects requires a sizable sample exposed to very high noise levels, but the number of residents living near airports where noise levels are so high – for example, exceeding 100 dBA from takeoffs or landings – is small.

In summary, noise exposure has been associated with reduced prenatal growth, and may be associated with reduced postnatal growth as well. Although most causal pathways between pollutant exposures and reduced growth remain obscure, noise-associated growth retardation may be related to the physiological components of the stress response to noise. If so, the study of chronic noise stress may have implications for understanding other chronic psychosocial stressors. An important advantage of the noise model for studying chronic stress effects is that environmental noise levels are commonly measured by government agencies and provide data about exposure. Such data are collected prospectively, and are independent of subject recall at the time of the growth study. In the study of noise exposure and physical growth, both the outcome and the stimulus can be measured objectively, whereas in many other studies either the ascertainment of the health outcome may be biased in some way, or the exposure itself is difficult to measure in the past.

Summary and conclusions

The study of growth and development among populations subject to exposure from environmental pollutants has newly emerged from traditional foci in auxology and epidemiology. In auxology, it represents an extension of the focus on urban–rural differences. New research can now build on traditional, large-scale urban–rural contrasts. Some research has already begun to focus on particular features, such as noise and lead, that differ between urban and rural areas. The traditional urban–rural study design involved a multidimensional contrast, and many factors were plausible explanations of the differences in growth. There has always been some uncertainty about which of the components of the urban environment actually influence growth, and the problem is compounded today by the presence of many new elements in urban environments.

Even when growth differs little between rural and urban areas, as in the United States (Hamill, Johnston & Lemeshow, 1972), the possibility remains that growth is affected by pollution. The above review of studies of individual pollutants suggests that they are detrimental to growth. The absence of an urban growth depression in well-off countries may be due to a balance between the positive influences of health care, nutrition, etc., and the unfavourable influences of pollutants and stress. In addition, some rural areas may be quite similar to urban ones in regard to some pollutants, or in providing the substrates for good growth. In other words, the extent of growth differences varies among urban–rural contrasts, and the difference between the urban and rural environments differs also. Only with controlled comparisons may the relationships between growth and pollutants be discovered, and the urban influences on growth identified. Specific data about environmental parameters is required.

In public health and epidemiology, the study of growth has emerged as an alternate means of evaluating the effects on community health of untoward pollutant exposures. The study of prenatal growth has long been a part of reproductive epidemiology. The clinical entities of low birthweight, congenital malformations, teratagenesis and spontaneous abortion have been the focus, but now there is nore attention to foetal growth generally, and to postnatal development also.

The study of growth in relation to pollutant exposures is a direct application of physical anthropology to human affairs. The application of growth studies as an early warning of environmental hazards to human well-being is based on two ideas:

1 that the effects of the environment on human biology and health exist on a continuous scale from well-being to pathology, and growth can be used to make fine distinctions along that scale; and
2 that literature reviewed in each of the case studies indicates that growth may be affected by many important pollutants even though significant pathology may not have been produced by these pollutants.

Growth studies represent a means of community health assessment which is an alternative to the traditional counts of mortality or of morbidity. These traditional health outcomes may not be useful to assess the impact of the environment on the health of younger segments of the population. In addition, these outcomes usually represent the endpoint of a slowly developing pathological process. In terms of assessing environmental impacts, there is a shorter lag-time for the presence of growth effects. Child growth is an objective measure of health, unbiased by controversy in the community. It is assessed without clinical surveys, which become complicated when used where there may be a variety of pollutants and many potential adverse health outcomes to be ascertained. An important new area in growth studies is behavioural teratology, the study of congenital behavioural defects and developmental delays (Riley & Vorhees, 1986). Recent research has shown that such defects are associated with numerous environmental features, most notably alcohol (Streissguth *et al.*, 1984), but also PCBs (Fein *et al.*, 1983) and lead (Davis & Svendsgaard, 1987). Behavioural teratologies may be especially sensitive at critical periods of neurological development.

In combining public health and epidemiology with auxology, a balance in measurement precision is reached: the objective measurement of growth can be balanced by the objective measurement of environment pollutants. Auxologists, familiar with the former, may produce more informative studies of urban–rural differences and of secular trends by describing precisely the environments of the contrasted populations. Environmental epidemiologists gain an objective measure of child health. It is one which summarizes innumerable physiological processes, is highly amenable to statistical analyses, and can be used to make distinctions along a continuum of health statuses.

As more and more communities are affected by pollution, the possibility of widespread health effects becomes greater. This is occurring within industrialized countries, and internationally, too. Until now, most attention has focused on the heavily industrialized areas of Europe, the United States and Japan. However, people in the Third World are now at increased risk, too. Third World Nations now receive the chemical wastes from industrialized ones where government regulation of chemical waste disposal has increased its cost. It has become cost-effective for some industries to transport their waste to nations without effective protective regulation or enforcement strategies. For Third World countries, the receipt of waste is a source of revenue, but it may be adding another health risk to an already staggering health burden.

Acknowledgement
I wish to thank Ralph DeNardo and Deirdre Sanders for their help in the preparation of this chapter.

References

Alexander, F. W. & Delves, H. T. (1981). Blood lead levels during pregnancy. *International Archives of Occupational and Environmental Health*, 48, 35–9.

Anderson, H. A. (1985). Evolution of environmental epidemiologic risk assessment. *Environmental Health Perspectives*, 62, 389–92.

Ando, Y. & Hattori, H. (1973). Statistical studies on the effects of intense noise during human fetal life. *Journal of Sound and Vibration*, 27, 101–10.

(1977). Effects of noise on human placental lactogen (HPL) levels in maternal plasma. *British Journal of Obstetrics and Gynecology*, 84, 115–18.

Antal, A., Timaru, J., Muncaci, E., Ardevan, E., Jonescu, A. & Sandulache, L. (1968). Les variations de la réactivité de l'organisme et de l'état de santé des enfants en rapport avec la pollution de l'air communal. *Atmospheric Environment*, 2, 383–92.

Baghurst, P. J., McMichael, A. J., Vimpani, G. V., Robertson, E. F., Clark, P. D. & Wigg, N. R. (1987). Determinants of blood lead concentrations of pregnant women living in Port Pirie and surrounding areas. *The Medical Journal of Australia*, 146, 69–73.

Baghurst, P., Oldfield, R., Wigg, N., McMichael, A., Robertson, E. & Vimpani, G. (1985). Some characteristics and correlates of blood lead in early childhood: preliminary results from the Port Pirie study. *Environmental Research*, 38, 24–30.

Barltrop, D. (1969). Transfer of lead to the human foetus. In *Mineral Metabolism in Paediatrics*, ed. D. Barltrop & W. L. Burland, pp. 135–51. Philadephia: F. A. Davis Co.

Beall, C. M. (1982). An historical perspective on studies of human growth and development in extreme environments. In *A History of American Physical Anthropology*, ed. F. Spencer, pp. 447–65. New York: Academic Press.

Bellinger, D. C., Leviton, A., Needleman, H. L., Waternaux, C & Rabinowitz, M. (1986). Low-level lead exposure and infant development in the first year. *Neurobehavior Toxicology and Teratology*, **8**, 151–61.

Bellinger, D. C., Leviton, A., Waternaux, C., Needleman, H. L. & Rabinowitz, M. (1987). Longitudinal analysis of prenatal and postnatal exposure and early cognitive development. *New England Journal of Medicine*, **316**, 1037–43.

Bellinger, D. C., Needleman, H. L., Leviton, A., Waternaux, C., Rabinowitz, M. B. & Nichols, M. L. (1984). Early sensory–motor development and prenatal exposure to lead. *Neurobehavior Toxicology and Teratology*, **6**, 387–402.

Bielicki, T. (1986). Physical growth as a measure of economic well-being of populations: the twentieth century. In *Human Growth. Volume 3, Postnatal Growth* (2nd edn), ed. F. Falkner & J. M. Tanner, pp. 283–305. New York: Plenum Press.

Bielicki, T. & Welon, Z. (1982). Growth data as indicators of social inequalities: the case of Poland. *Yearbook of Physical Anthropology*, **25**, 153–67.

Bithoney, W. G. (1986). Elevated lead levels in children with nonorganic failure to thrive. *Pediatrics*, **78**, 891–5.

Bloom, A. D. (1981). *Guidelines for Studies of Human Populations Exposed to Mutagenic and Reproductive Hazards*. White Plains, NY: March of Dimes Birth Defects Foundation.

Boas, F. (1912). Changes in bodily form of descendants of immigrants. *American Anthropologist*, **14**, 530–62.

Campbell, D. T. & Stanley, J. C. (1963). *Experimental and Quasi-experimental Designs for Research*. Chicago: Rand McNally College Publishing Company.

Cantrell, R. W. (1974). Prolonged exposure to intermittent noise: audiometric biochemical, motor, psychological and sleep effects. *Laryngoscope*, **84**, 1–55 (10, part 2, supplement 1).

Coblentz, A. & Martel, A. (1986). Effects of fetal exposition to noise on the birthweight of children. *American Journal of Physical Anthropology* (abstract), **69**, 188.

Cohen, A. (1968). Noise effects on health, productivity and well-being. *Transactions of the New York Academy of Science*, **30**, 910–18.

(1973). Industrial noise and medical absence, and accident record data on exposed workers. In *Proceedings of the International Congress on Noise as a Public Health Problem*, ed. United States Enviromental Protection Agency, pp. 431–9. Washington: US Government Printing Office.

Cohen, S., Krantz, D. S., Evans, G. W. & Stokola, D. (1981). Cardiovascular and behavioral effects of community noise. *American Scientist*, **69**, 528–35.

Colucci, A. V., Hammer, D. I., Williams, M. E., Hinners, T. A., Pinkerton, C., Kent, J. L. & Love, G. J. (1973). Pollutant burdens and biological response. *Archives of Environmental Health*, **27**, 151–4.

Damstra, T. (1977). Toxicological properties of lead. *Environmental Health Perspectives*, **19**, 297–307.

Davis, J. M. & Svendsgaard, D. J. (1987). Lead and child development. *Nature*, **329**, 297–300.

Dietrich, K. N., Krafft, K. M., Beir, M., Succop, P. A., Berger, O. & Bornschein, R. L. (1986). Early effects of fetal lead exposure: neurobehavioral findings at 6 months. *International Journal of Biosocial Research*, **8**, 2–19.

Dietrich, K. N., Krafft, K. M., Bornschein, R. L., Hammond, P. B., Berger, O., Succop, P. A. & Beir, M. (1987). Low-level fetal lead exposure effect on neurobehavioral development in early infancy. *Pediatrics*, **80**, 721–30.

Eisenbud, M. (1978). *Environment, Technology, and Health.* New York: New York University Press.

Ernhart, C. B., Morrow-Tlucak, M., Marler, M. R. & Wolf, A. W. (1987). Low level lead exposure in the prenatal and early preschool periods: early preschool development. *Neurotoxicology and Teratology*, **9**, 259–70.

Ernhart, C. B., Wolf, A. W., Kennard, M. J., Erhard, P., Filipovich, H. F. & Sokol, R. J. (1986). Intrauterine exposure to low levels of lead: the status of the neonate. *Archives of Environmental Health*, **41**, 287–91.

Eveleth, P. (1986). Population differences in growth: environmental factors. In *Human Growth, Volume 3, Postnatal growth* (2nd edn), ed. F. Falkner & J. M. Tanner, pp. 221–39. New York: Plenum Press.

Eveleth, P. & Tanner, J. M. (1976). *Worldwide Variation in Human Growth.* Cambridge: Cambridge University Press.

Fahim, M. S., Fahim, Z. & Hall, D. G. (1976). Effects of subtoxic lead levels on pregnant women in the state of Missouri. *Research Communications in Chemical Pathology and Pharmacology*, **13**, 309–31.

Fein, G. G., Schwartz, P. M., Jacobson, S. W. & Jacobson, J. L. (1983). Environmental toxins and behavioral developmental: a new role for psychological research. *American Psychologist*, **38**, 1188–97.

Fein, G. G., Jacobson, J. L., Jacobson, S. W., Schwartz, P. M. & Dowler, J. K. (1984). Prenatal exposure to polychlorinated biphenyls: effects on birth size and gestational age. *Journal of Pediatrics*, **105**, 315–20.

Frisancho, A. R. (1979). *Human Adaptation.* St Louis: Mosby.

Fujisawa, H. & Fujiwara, B. (1972). On the influence which PCB exercises on the development of the child. *Bulletin of the Faculty of Liberal Arts, Nagasaki University, Natural Science*, **13**, 15–21.

Funatsu, I., Yamashita, F., Yoshikane, T., Funatsu, T., Ito, Y., Tsugawa, S., Hayashi, M., Kato, T., Yakushiji, M., Okamoto, G., Arima, A., Adachi, N., Takahashi, K., Miyahara, M., Tashiro, Y., Shimomura, M., Yamasaki, S., Arima, T., Kuno, T., Ide, H. & Ide, I. (1971). A chlorobiphenyl induced fetopathy. *Fukuoka Acta Medica*, **62**, 139–49.

Gershanik, J. J., Brooks, G. G. & Little, J. A. (1974). Blood lead values in pregnant women and their offspring. *American Journal of Obstetrics and Gynecology*, **119**, 508–11.

Goldman, L. R., Paigen, B., Magnant, M. M. & Highland, J. H. (1985). Low birth weight, prematurity and birth defects in children living near the hazardous waste site, Love Canal. *Hazardous Waste and Hazardous Materials*, **2**, 209–23.

Habercam, J. W., Keil, J. E., Reigart, J. R. & Croft, H. W. (1974). Lead content of human blood, hair, and deciduous teeth: correlation with environmental factors and growth. *Journal of Dental Research*, **53**, 1160–3.

Hamill, P. V., Johnston, F. E. & Lemeshow, S. (1972). *Height and Weight of Children: Socioeconomic Status.* United States Vital and Health Statistics, Series 11, no. 119. Washington, DC: US Department of Health Education and Welfare, Publication No. HSM 73–1601.

Hara, I. (1985). Health status and PCBs in blood of workers exposed to PCBs and of their children. *Environmental Health Perspectives*, **59**, 85–90.

Hartikainen-Sorri, A. -L., Sorri, M., Anttonen, H. P., Tuimala, R. & Laara, E. (1988). Occupational noise exposure during pregnancy: a case control study. *Occupational and Environmental Health*, **60**, 279–83.

Houk, V. N. (1982). Determining the impacts on human health attributable to hazardous waste sites. In *Risk Assessment at Hazardous Waste Sites*, ed. F. A. Long & G. E. Schweitzer, pp. 21–32. Washington: American Chemical Society.

Huel, G., Boudene, C. & Ibrahim, M. A. (1981). Cadmium and lead content of maternal and newborn hair: relationship to parity, birth weight, and hypertension. *Archives of Environmental Health*, **36**, 221–7.

Hunter, J. (1986). The distribution of lead. In *Lead Toxicity: History and Environmental Impact*, ed. R. Landsdown & W. Yule, pp. 96–126. Baltimore: The Johns Hopkins University Press.

Jacobson, J. L., Fein, G. G., Jacobson, S. W., Schwartz, P. M. & Dowler, J. K. (1984). The transfer of polychlorinated biphenyls (PCBs) and polybrominated biphenyls (PBBs) across the human placenta and into maternal milk. *American Journal of Public Health*, **74**, 378–9.

Johnson, N. E. & Tenuta, K. (1979). Diets and lead blood levels of children who practice pica. *Environmental Research*, **18**, 369–76.

Johnston, F. E. (1983). The uses of anthropometry. *Acta Medica Auxologica*, **15**, 69–74.

Johnston, F. E., Low, S. M., de Baessa, Y. & McVean, R. B. (1985). Growth status of disadvantaged urban Guatemalan children of a resettled community. *American Journal of Physical Anthropology*, **68**, 215–24.

Kaplan, B. A. (1954). Environment and human plasticity. *American Anthropologist*, **56**, 780–99.

Karagodina, I. L., Soldatkina, S. A., Vinokur, I. L. & Klimukhin, A. A. (1969). Effect of aircraft noise on the population near airports. *Hygiene Sanitation*, **34**, 182–7.

Kawai, M., Okamoto, Y. & Katagiri, Y. (1987). Variation of blood-levels with age in childhood. *International Archives of Occupational and Environmental Health*, **59**, 91–4.

Knipschild, P. (1977a). Medical effects of aircraft noise: general practice survey. *International Archives of Occupational and Environmental Health*, **40**, 191–6.

(1977b). Medical effects of aircraft noise: community cardiovascular survey. *International Archives of Occupational and Environmental Health*, **40**, 185–90.

Knipschild, P., Meijer, H. & Salle, H. (1981). Aircraft noise and birthweight. *International Archives of Occupational and Environmental Health*, **48**, 131–6.

Knipschild, P. & Oudschoorn, N. (1977). Medical effects of aircraft noise: drug

112   L. M. Schell

survey. *International Archives of Occupational and Environmental Health*, **40**, 197–200.

Kryter, K. D. (1985). *The Effects of Noise on Man*. (2nd edn). New York: Academic Press.

(1987). Environmental aircraft noise and social factors in stress-related health disorders. In *Noise Control in Industry: Internoise 87*, Vol. *II*, pp. 949–52. Beijing: Acoustical Society of China.

Kuwabara, K., Yakushiji, T., Watanabe, I., Yoshida, S., Koyama, K., Kunita, N. & Hara, I. (1978). Relationship between breast feeding and PCB residues in blood of the children whose mothers were occupationally exposed to PCBs. *International Archives of Occupational and Environmental Health*, **41**, 189–97.

Landrigan, P. J. (1983). Epidemiologic approaches to persons with exposures to waste chemicals. *Environmental Health Perspective*, **48**, 93–7.

Lasker, G. W. (1969). Human biological adaptability. *Science*, **166**, 1480–6.

Lauwers, M.-C., Hauspie, R., Susanne, C. & Verheyden, J. (1986). Comparison of biometric data of children with high and low levels of lead in the blood. *American Journal of Physical Anthropology*, **69**, 107–16.

Lauwers, M.-C., Hauspie, R., Thiessen, L., Verduyn, G. & Susanne, C. (1984). Analysis of growth in height and weight of Belgian children in relation to lead levels in the air. In *Human Growth and Development*, ed. J. Borms, R. Hauspie, A. Sand, C. Susanne & M. Hebbelinck, pp. 115–24. New York: Plenum Press.

Lowrance, W. (ed.) (1981). *Assessment of Health Effects at Chemical Waste Sites*. New York: Rockefeller University.

Lubin, A. H., Caffo, A. L. & Reece, R. (1978). A longitudinal study of interaction between environmental lead and blood lead concentrations during pregnancy, at delivery, and the first 6 months of life. *Pediatric Research*, **12**, 425.

MacMahon, B. & Pugh, T. F. (1970). *Epidemiology: principles and methods*. Boston: Little, Brown & Co.

McMichael, A. J., Baghurst, P. A., Wigg, N. R., Vimpanni, G. V., Robertson, E. F. & Roberts, R. J. (1988). Port Pirie cohort study: environmental exposure to lead and children's abilities at the age of four years. *New England Journal of Medicine*, **319**, 468–75.

McMichael, A. J., Vimpani, G. V., Robertson, E. F., Baghurst, P. A. & Clark, P. D. (1986). The Port Pirie cohort study: maternal boood lead and pregnancy outcome. *Journal of Epidemiology and Community Health*, **40**, 18–25.

Mahaffey, K. R. (ed.) (1985). *Dietary and Environmental Lead: human health effects*. New York: Elsevier.

Mahaffey, K. R., Annest, J. L., Roberts, M. S. & Murphy, R. S. (1982). National estimates of blood lead levels: United States, 1976–1980. *The New England Journal of Medicine*, **307**, 573–9.

Malina, R. M., Himes, J. H., Stepick, C. D., Lopez, F. G. & Bouchang, P. H. (1981). Growth of rural and urban children in the Valley of Oaxaca, Mexico. *American Journal of Physical Anthropology*, **55**, 269–80.

Marecek, J., Shapiro, I. M., Burke, A., Katz, S. H. & Hediger, M. L. (1983).

Low-level lead exposure in childhood influences neuropsychological performance. *Archives of Environmental Health*, **38**, 355–9.

Masuda, Y. & Yoshimura, H. (1984). Polychlorinated biphenyls and dibenzofurans in patients with Yusho and their toxicological significance: a review. *American Journal of Industrial Medicine*, **5**, 31–44.

Mikusek, J. (1976). Developmental age and growth of girls from regions with high atmospheric air pollution in Silesia. *Roczniki Panstwowego Zakludu Higieny*, **27**, 473–81.

Miller, R. W. (1985). Congenital PCB poisoning: a reevaluation. *Environmental Health Perspectives*, **60**, 211–14.

Mooty, J., Ferrand, C. F. & Harris, P. (1975). Relationship of diet to lead poisoning in children. *Pediatrics*, **55**, 636–9.

Needleman, H. L. (1983). Lead at low dose and the behavior of children. *Acta Psychiatrica Scandinavica*, **67**(suppl. 303), 26–37.

(1987) Letter to the editor. *Archives of Environmental Health*, **42**, 242.

Needleman, H. L., Davidson, I., Sewell, E. M. & Shapiro, I. M. (1974). Subclinical lead exposure in Philadelphia schoolchildren. *New England Journal of Medicine*, **290**, 245–8.

Needleman, H. L., Gunnoe, C., Leviton, A., Reed, R., Peresie, H., Maher, C. & Barrett, P. (1979). Deficits in psychologic and classroom performance of children with elevated dentine lead levels. *New England Journal of Medicine*, **300**, 689–95.

Nordstrom, S., Beckman, S. & Nordenson, I. (1978). Occupational and environmental risks in and around a smelter in northern Sweden. *Hereditas*, **88**, 43–6.

Nriagu, J. O. (1983). *Lead and Lead Poisoning in Antiquity*. New York: John Wiley.

Okumura, M. (1984). Past and current medical states of Yusho patients. *American Journal of Industrial Medicine*, **5**, 13–18.

Paigen, B., Goldman, L. R., Highland, J. H., Magnant, M. M. & Steegmann, A. T. Jr (1985). Prevalence of health problems in children living near Love Canal. *Hazardous Waste and Hazardous Materials*, **2**, 23–43.

Paigen, B., Goldman, L. R., Magnant, M. M., Highland, J. H. & Steegmann, A. T. Jr (1987). Growth of children living near the hazardous waste site, Love Canal. *Human Biology*, **59**, 489–508.

Pollitt, E. (1987). Effects of iron deficiency on mental development: methodological considerations and substantive findings. In *Nutritional Anthropology*, ed. F. E. Johnston, pp. 225–54. New York: Alan R. Liss.

Prichard, J. C. (1813). *Researches into the Physical History of Man*; 1973 ed. G. W. Stocking Jr. Chicago: University of Chicago Press.

Quah, R. F., Stark, A. D., Meigs, J. W. & DeLouise, E. R. (1982). Children's blood lead levels in New Haven: a population-based demographic profile. *Environmental Health Perspectives*, **44**, 159–64.

Rabinowitz, M., Leviton, A. & Needleman, H. (1984). Variability of blood lead concentrations during infancy. *Archives of Environmental Health*, **39**, 74–7.

Rajegowda, B. K., Glass, L. & Evans, H. E. (1972). Lead concentrations in the newborn infant. *Journal of Pediatrics*, **80**, 116–18.

114   L. M. Schell

Riley, E. P. & Vorhees, C. V. (eds) (1986). *Handbook of Behavioral Teratology*. New York: Academic Press.

Rogan, W. J., Gladen, B. C., McKinney, J. D., Carreras, N., Hardy, P., Thullen, J., Tinglestad, J. & Tully, M. (1986). Neonatal effects of transplacental exposure to PCBs and DDE. *Journal of Pediatrics*, 109, 335–41.

Rogan, W. J., Gladen, B. C. & Wilcox, A. J. (1985). Potential reproductive and postnatal morbidity from exposure to polychlorinated biphenyls: epidemiologic considerations. *Environmental Health Perspectives*, 60, 233–9.

Roncevic, N., Pavkov, S., Galetin-Smith, R., Vukavic, T., Vojinovic, M. & Djordjevic, M. (1987). Serum concentrations of organochlorine compounds during pregnancy and the newborn. *Bulletin of Environmental Contamination and Toxicology*, 38, 117–24.

Schell, L. M. (1981). Environmental noise and human prenatal growth. *American Journal of Physical Anthropology*, 56, 63–70

   (1984). Auxological epidemiology and the determination of the effects of noise on health. In *Genetic and Environmental Factors during the Growth Period*, ed. C. Susanne, pp. 209–19. New York: Plenum Press.

   (1986). Community health assessment through physical anthropology: auxological epidemiology. *Human Organization*, 45, 321–7.

   (1988). Cities and human health. In *Urban Life*. (2nd edn), ed. G. Gmelch & W. P. Zenner, pp. 18–35. Prospect Heights, Illinois: Waveland Press.

Schell, L. M. & Hodges, D. C. (1985). Longitudinal study of growth status and airport noise exposure. *American Journal of Physical Anthropology*, 66, 383–9.

   (1987). Airport noise and human physical development. In *Noise Control in Industry: Internoise 87, Vol. II*, pp. 957–60. Beijing: Acoustical Society of China.

Schell, L. M. & Lieberman, L. S. (1981). Noise stress and cancer. In *Stress and Cancer*, ed. K. Bammer & B. N. Newberry, pp. 202–31. Toronto: Hogrefe.

Schell, L. M. & Norelli, R. J. (1983). Airport noise and the postnatal growth of children. *American Journal of Physical Anthropology*, 61, 473–82.

Schmidt, P. & Dolgner, R. (1977). Interpretation of some results of studies in school-children living in areas with different levels of air pollution. *Zentralblatt für Bakteriologie, Parasitenkunde, Infektionskrankheiten und Hygiene, I bt. Orig. B*, 165, 539–47.

Schwartz. J., Angle, C. & Pitcher, H. (1986). Relationship between childhood blood lead levels and stature. *Pediatrics*, 77, 281–8.

Stark, A. D., Standfast, S. D. & Huffaker, R. H. (1985). Hazardous waste sites. *Postgraduate Medicine*, 77, 13–19.

Streissguth, A. P., Martin, D. C., Barr, H. M., Sandman, B. M. Kirchner, G. L. & Darby, B. L. (1984). Intrauterine alcohol and nicotine exposure: attention and reaction time in 4-year-old children. *Developmental Psychology*, 20, 533–41.

Takahashi, I. & Kyo, S. (1968). Studies on the differences in adaptabilities to the noise environment in sexes and growing processes. *Journal of the Anthropological Society of Nippon*, 76, 34–51.

Taki, I., Hisanaga, S. & Amagase, Y. (1969). Report on Yusho (chlorobiphenyls poisoning): pregnant women and their fetuses. *Fukuoka Acta Medica*, 60, 471–4.

Tanner, J. M. (1966). Growth and physique in different populations of mankind. In *The Biology of Human Adaptability*, ed. P. T. Baker & J. S. Weiner, pp. 45–66. Oxford: Clarendon Press.

(1979). A concise history of growth studies from Buffon to Boas. In *Human Growth. Volume 3, Nutrition and Neurophysiology*, ed. F. Falkner & J. M. Tanner, pp. 515–93. New York: Plenum Press.

(1981). *A History of the Study of Human Growth*. Cambridge: Cambridge University Press.

(1986). Growth as a mirror of the condition of society: secular trends and class distinctions. In *Human Growth: a multidisciplinary review*, ed. A. Demirjian, pp. 3–34. Basingstoke, Hants: Taylor & Francis.

Taylor, P. R., Lawrence, C. E., Hwang, H.-L. & Paulson, A. S. (1984). Polychlorinated biphenyls: influence on birthweight and gestation. *American Journal of Public Health*, 1153–4.

Taylor, P. R., Stelma, J. & Lawrence, C. E. (1989). The relation of polychlorinated biphenyls to birthweight and gestational age in the offspring of occupationally exposed mothers. *American Journal of Epidemiology*, **129**, 395–406.

Thielebeule Von., U., Pelech, L., Grosser, P. J. & Horn K. (1980). Height and bone age of school children in areas of different concentrations of air pollution. *Zeitschrift für die Gesamte Hygiene und Ihre Grenzgebiete*, **26**, 771–3.

Topinard, P. (1890). *Anthropology*. London: John Chapman.

Vianna, N. J. (1980). Adverse pregnancy outcomes – potential endpoints of human toxicity in the Love Canal: preliminary results. In *Embryonic and Fetal Death*, ed. I. H. Porter & E. B. Hook, pp. 165–8. New York: Academic Press.

Vianna, N. J. & Polan, A. K. (1984). Incidence of low birth weight among Love Canal residents. *Science*, **226**, 1217–19.

Waldbott, G. L. (1978). *Health Effects of Environmental Pollutants*, (2nd edn). St Louis: Mosby.

Watkins, G., Tarnopolsky, A. & Jenkins, L. M. (1981). Aircraft noise and mental health: II. Use of medicines and health care services. *Psychological Medicine*, **11**, 155–68.

Welch, B. L. & Welch, A. M. (1970). *Physiological Effects of Noise*. New York: Plenum Press.

Westman, J. C. & Walters, J. R. (1981). Noise and stress: a comprehensive approach. *Environmental Health Perspectives*, **41**, 291–309.

Wibberley, D. G., Khera, A. K., Edwards, J. H. & Rushton, D. I. (1977). Lead levels in human placentae from normal and malformed births. *Journal of Medical Genetics*, **14**, 339–45.

Williams, L. M., Spence, A. & Tideman, S. C. (1977). Implications of the observed effect of air pollution on birth weight. *Social Biology*, **24**, 1–9.

World Health Organization (1981). *Basic Documents* (31st edn). Geneva: World Health Organization.

Yakushiji, T., Watanabe, I., Kuwabara, K., Tanaka, R., Kashimoto, T. & Kunita, N. (1984). Postnatal transfer of PCBs from exposed mothers to their babies: influence of breastfeeding. *Archives of Environmental Health*, **39**, 368–75.

116   *L. M. Schell*

Yamaguchi, A., Yoshimura, T. & Kuratsune, M. (1971). A survey on pregnant women having consumed rice oil contaminated with chlorobiphenyls and their babies. *Fukuoka Acta Medica*, **62**, 117–22.

Yamashita, F. & Hayashi, M. (1985). Fetal PCB syndrome: clinical features, intrauterine growth retardation and possible alteration in calcium metabolism. *Environmental Health Perspectives*, **59**, 41–5.

Yoshimura, T. (1971). A case control study on growth of school children with Yusho. *Fukuoka Acta Medica*, **62**, 109–16.

(1974). Epidemiological study on Yusho babies born to mothers who had consumed oil contaminated by PCB. *Fukuoka Acta Medica*, **65**, 74–80.

Yoshimura, T. & Ikeda, M. (1978). Growth of school children with polychlorinated biphenyl poisoning or Yusho. *Environmental Research*, **17**, 416–25

5 Human physiological adaptation to high-altitude environments

LAWRENCE P. GREKSA

The evolutionary adaptation of populations to their environments, each with its unique set of interrelated stresses, has resulted in considerable human biological variation (Baker, 1988). Assessing the magnitude of this variation and determining its causes and functional significance is a central concern for the discipline of human biology. In addition, human biological variation has important practical implications, and ones which are sometimes overlooked, for some aspects of the implementation of public health programmes. As one example, assuming that milk was equally nutritious for all individuals, European nations frequently included dehydrated milk in their food supplementation programmes to African and Asian countries after World War II. However, many Africans and Asians possess significantly lower concentrations than Europeans of the enzyme lactase, which is required for the digestion of the lactose in milk. As a result, the ingestion of milk irritated the intestine and thus frequently resulted in diarrhoea (Baker, 1988). In other words, a lack of knowledge about biological variation led to the development of an inappropriate, and in fact highly maladaptive, programme.

One successful research strategy developed and utilized by human biologists for examining biological variation and for assessing its functional significance involves measuring human biological responses to severe environmental stresses (Baker, 1988). A particularly fruitful area of such research has involved determining the biological consequences of exposure to the decrease in oxygen availability which occurs in high-altitude environments. The aim of this paper is to review the considerable research which has been conducted on the physiological responses to lifelong residence in high-altitude environments. Besides providing an updated review for human biologists, this review illustrates the potential magnitude of human biological variation for those interested in developing science policy. Prior to reviewing this literature, some key terms and concepts will first be presented.

Background

High-altitude environments are multi-stress environments, but the primary stress is a reduction in atmospheric pressure and, as a result, a reduction in the atmospheric partial pressure of oxygen (P_{O_2}), or hypobaric hypoxia (Baker, 1969). Expressed differently, there are fewer oxygen molecules per unit of inspired air at high altitudes than at low altitudes. Altitudes above 2500 m, or the altitude at which the decrease in atmospheric P_{O_2} is physiologically significant in most low-to-high altitude migrants, are generally considered to be high-altitude environments. Not surprisingly, a basic aim of high-altitude research is to determine the mechanisms which operate to provide a continuous input of adequate amounts of oxygen to metabolic tissue, despite residence within an oxygen-deficient environment.

Unlike other environmental stresses studied by human biologists, behavioural responses are generally ineffective in ameliorating the effects of hypobaric hypoxia. As a result, any adaptive responses to this stress much occur within the oxygen transport system, or the biological system involved in the uptake, transport and utilization of oxygen. It will therefore be useful briefly to describe this system. The initial step in the oxygen transport system is pulmonary ventilation, or the movement of air into and out of the lungs. A proportion of the inspired air diffuses across the alveolar membranes into arterial blood, or from an area of higher to lower P_{O_2}. Most of the diffused oxygen attaches to haemoglobin, one component of red blood cells. Red blood cells are then pumped by the heart to the working tissues. Once red blood cells reach the working tissues, a proportion of the oxygen which they carry dissociates from haemoglobin and diffuses across cellular membranes, once again from an area of higher to lower P_{O_2}. At the tissue level, the oxygen is used to support oxidation.

Comprehensive studies of high-altitude adaptation were initiated in Peru in 1928 by Carlos Monge M. As part of his efforts, Monge organized the Institute of Andean Biology at the University of San Marcos in Lima, Peru. With the acceptance of evolutionary theory as the primary paradigm of the biological sciences in the 1950s, there was an increased interest in the nature of the relationship between the environment and human biology (Little, 1982). The International Council of Scientific Unions established the International Biological Programme (IBP), one component of which was the Human Adaptability Project. Studies of high-altitude adaptation were sponsored by this section in all of the major high-altitude ecozones between 1967 and 1972 (Baker, 1978).

With this background, it will now be possible to review the research on the physiological responses of native highlanders to their hypobaric hypoxic environment. An emphasis will be placed on post-IBP studies.

Lung volumes

The capacity of the lungs to move air is partially determined by total lung capacity, two components of which are vital capacity and residual volume. Vital capacity is the total volume of air that can be voluntarily moved in one breath, from full inspiration to maximal expiration. If the expiration is as rapid and forceful as possible, it is termed a forced vital capacity (FVC). Residual volume is that portion of total lung capacity which remains after a total expiration. Most studies at high altitude have focused on FVC.

The first comprehensive studies of lung volumes were conducted in Peru by Alberto Hurtado of the Institute of Andean Biology. Hurtado (1932, 1964) demonstrated that male Quechua Indian highlanders possessed significantly larger total lung capacities than lowlanders of the same stature, primarily as a result of enhanced residual volumes and secondarily due to increased vital capacities. Hurtado argued that an increased vital capacity was adaptive because it ensured adequate ventilation of the alveoli. Hurtado (1932) also conducted anatomical investigations which indicated that highlanders had dilated lung capillaries, which he argued was adaptive since it increased the potential for diffusion from the alveoli into the circulating blood.

A number of IBP reseachers examined the effect of high altitudes on FVC. In perhaps the most influential study, Frisancho (1969; Frisancho & Baker, 1970) found that highland Quechua Indians living in Nuñoa, Peru (4000 m) were shorter than either lowland Peruvians or individuals of European ancestry residing at low altitude in the United States, but nevertheless had significantly larger FVCs. In a later study, Frisancho, Velasquez & Sanchez (1973a) found that FVC was similar between highland natives and adult males who had migrated to altitude as children but that both groups had significantly larger FVCs than men who had migrated as adults. Enlarged lung volumes among highlanders were also reported from Ethiopia (Clegg *et al.*, 1972) and Peru (Boyce *et al.*, 1974) but not in the Tien Shan mountains (Mirrakhimov, 1978) or in one study in the Andes (Hoff, 1974).

The Institute of Andean Biology researchers hypothesized that Andean highlanders had adapted genetically to their hypobaric hypoxic environment (Monge, 1948). Based on the studies described above, as well as studies of other components of the oxygen transport system (to be described shortly), Frisancho (1976) hypothesized that developmental adaptation played a more important role than genetic adaptation. In other words, he hypothesized that exposure to hypobaric hypoxia during growth and development, with late childhood and adolescence being particularly critical periods, resulted in permanent beneficial changes in the structure and function of the cardiorespiratory system. In particular, he argued that cardiorespiratory system organs displayed an accelerated development relative to stature in highlanders during late childhood and especially adolescence. Some of the data collected by other IBP researchers were consistent with this hypothesis (for example, Boyce *et al.*, 1974; Lahiri *et al.*, 1976), but studies in Ethiopia (Harrison *et al.*, 1969) and in Peru (Boyce *et al.*, 1974) found that enhanced lung volumes could be attained by individuals who migrated to high altitude as adults.

Frisancho's hypothesis, which was primarily based on Quechua growth patterns, was, and still is, widely accepted. As a result, much of the post-IBP research on lung volumes in native highlanders has focused on the magnitude and patterns of development of FVC. The Multinational Andean Genetic and Health Program found a significant positive relationship between altitude and FVC in Aymara Indians residing in Chile and Bolivia (Mueller *et al.*, 1978). However, the magnitude of the increase was less than predicted on the basis of Frisancho's studies of Nuñoa Quechua Indians. Similar deviations from expectations were reported by Beall (1984; Beall, Strohl & Brittenham, 1983), who found that mean FVC in subjects of Tibetan ancestry was significantly smaller than in Nuñoa Quechua; that FVC was not greatly, if at all, enhanced above sea level values in highland Tibetans; and that Tibetans did not appear to exhibit an accelerated development of FVC relative to stature during late childhood and adolescence. Hackett *et al.* (1980) also reported that FVC was not enhanced above sea level values in highland men of Tibetan ancestry. In addition, highland urban youths of Aymara ancestry residing in La Paz, Bolivia, have significantly smaller FVCs and different patterns of development of FVC than Nuñoa Quechua (Greksa *et al.*, 1987).

Thus, virtually all studies of native highlanders conducted since the conclusion of the IBP have detected considerable population variability in the magnitude and pattern of development of FVC, as demonstrated in Fig. 5.1. It is worth noting that examination of Fig. 5.1 suggests that the hypoxic response of Quechua Indians, rather than being normative, may

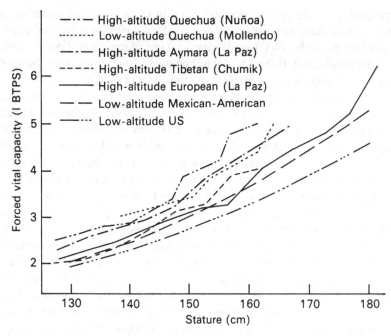

Fig. 5.1. Forced vital capacity in selected populations (males). (Adapted from Greksa *et al.*, 1988.)

in fact be atypical (Greksa & Beall, 1989). For example, *low-altitude* Quechua children from Mollendo, Peru have FVCs as large as those of *high-altitude* Aymara children from La Paz (Fig. 5.1).

An important goal of high-altitude research is to determine if the population differences in FVC demonstrated in Fig. 5.1 reflect either differential levels of adaptation (that is, population differences in the degree to which the effects of hypoxic stress are ameliorated), different patterns of adaptation (that is, differences between populations in their specific physiological responses to hypobaric hypoxia), or both. One problem in resolving this issue is that most highlanders have a long history of residence at high altitudes, making genetic adaptation a possibility. One strategy for controlling for genetic factors would be to measure the physiological effects of chronic hypoxia in a control population of highlanders who have a short history of residence at high altitude and are therefore unlikely to have adapted through genetic processes. It could reasonably be inferred that populations with a long history of residence at high altitude have not responded genetically if their biological characteristics (for example, FVC) are similar to those of the control population. On the other hand, if their biological characteristics differ from those of the control population, genetic responses are a possibility.

Individuals of European ancestry residing in highland Bolivia fulfil the requirements of a control population (Greksa, 1988). Studies have been conducted on children of European ancestry who were born and raised at high altitudes (European sedentes) and children of European ancestry who were born at low altitudes and migrated to high altitude at a later age (European migrants). European sedentes tend to have smaller FVCs than either Andean or Himalayan highland natives of the same stature (Fig. 5.1; Greksa et al., 1988). As expected, European migrants have significantly smaller FVCs than European sedentes (Greksa, 1988). Comparison of the European sedentes and migrants with appropriate lowland controls suggested that lifelong exposure to chronic hypoxia results in a statistically significant increase in average FVC of between 150 and 400 ml (Greksa et al., 1987; Greksa, 1988). Since growth at high altitude results in small to moderate average increases in FVC in the control population of European children, the apparent lack of a hypoxic effect in Tibetans and the presence of an even larger effect among Andean Indians suggest that genetic factors could be operating in both groups. However, since the variation in mean FVC between highland groups (Fig. 5.1) is no greater than found between different lowland ethnic groups (Hsu et al., 1979), such an interpretation seems unwarranted at this time.

Finally, quantitative assessments of the developmental adaptation hypothesis in the European sedente sample did not provide much support for this hypothesis (Greksa et al., 1988). On the other hand, the differences, or residuals, between the actual FVC of each European migrant child and the FVC of a European highland sedente of the same stature were significantly and negatively related to age of migration to high altitude (Fig. 5.2; Greksa, 1988). In other words, migrants tend to become more similar to sedentes with increasing exposure to chronic hypoxia, which is consistent with the general expectations of developmental adaptation. However, the magnitude of the increase in FVC associated with a given length of exposure to chronic hypoxia appears to be similar at all periods of growth and not greatest during adolescence, as predicted by the developmental adaptation hypothesis (Greksa, 1988). In addition, the data on European sedentes, as well as other highland

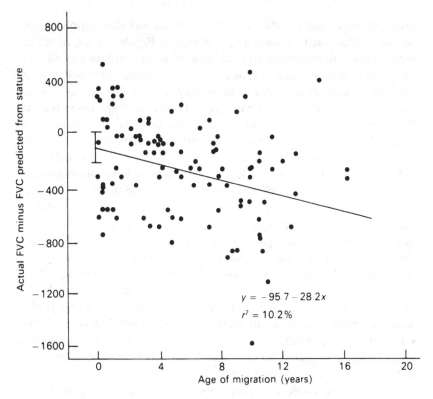

Fig. 5.2. Relationship between FVC residuals and age of migration to high altitude in European migrants. The FVC residuals measure the differences between the actual FVC of each migrant child and the estimated FVC for a European highland native of the same stature. The brackets represent the 95 % confidence interval for a predicted FVC residual at an age of migration of 0 years. (Adapted from Greksa, 1988.)

groups, suggests that the enhanced lung volumes of highlanders are already established by early childhood and are merely maintained during late childhood and adolescence (cf. Fig. 5.1).

Ventilation

Given the larger lung volumes of highlanders, it is not surprising that the volume of air inspired per minute [V_E (litre/min)] is significantly larger in highlanders than in lowlanders (Velasquez, 1976). Ventilation is primarily regulated by chemoreceptors which monitor various blood characteristics, such as arterial P_{O_2}. If arterial P_{O_2} decreases, as occurs with exposure to hypobaric hypoxia, an increase in ventilation is initiated by chemoreceptors. The magnitude of the increase in ventilation in lowlanders at high altitudes is frequently used to define the normal ventilatory response to hypobaric hypoxia. Native highlanders tend to have smaller V_Es than lowlanders under the same hypoxic stress, indicating they have a reduced chemosensitivity to hypoxia, or what has generally been referred to as a blunted hypoxic drive (Lahiri, 1984).

V_E is about 10 % lower in highlanders than in low-to-high altitude migrants at rest and the difference increases to 20–25 % during maximal exercise (Cerretelli, 1980). Since highlanders require less ventilation (and therefore less respiratory work) to move the same amount of oxygen as lowland migrants, this pattern is generally considered to be adaptive (Hurtado, 1964). A blunted hypoxic drive has consistently been found in the Andes (Cudkowicz, Spielvogel & Zubieta, 1972; Lahiri, 1984). Whether the same is true for the Himalayas is not clear (Lahiri, 1984), although there is increasing evidence that Tibetan natives may not exhibit a strongly blunted hypoxic drive (Hackett *et al.*, 1980; Sun *et al.*, 1988). Given that other differences between Andean and Himalayan natives appear to exist, a finding that they also differ in their ventilatory response to hypoxia would not be surprising.

In studies of Andean highlanders, Lahiri and colleagues (1978) demonstrated that a blunted hypoxic drive is not present at birth. Instead, it develops gradually after birth. It begins to be expressed in some individuals during adolescence and is expressed in all individuals by early adulthood (Lahiri *et al.*, 1976). This pattern of increasing insensitivity to hypoxia during growth and development is associated with an increase in vital capacity. Studies of migrant children and adults also indicated that the blunted hypoxic drive of highlanders is acquired through exposure to hypobaric hypoxia during the period of growth and development (Lahiri *et al.*, 1976). Similar findings were reported by Weil *et al.* (1971).

One situation during which the ventilatory response of highland natives tends to be less blunted than normally is during pregnancy. Since the oxygen requirements of both the mother and the foetus must be met, the potential for hypoxic stress is high during pregnancy (Haas, 1980). Moore and colleagues (1982a, 1986) demonstrated that the primary adaptive response to pregnancy by highland females of European ancestry residing in Leadville, Colorado (3100 m), and by highland Andean females (4300 m) was an increase in V_E. As a result, oxygen delivery

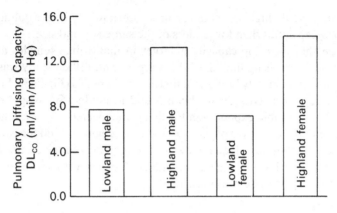

Fig. 5.3. Pulmonary diffusing capacity in young lowland and highland children. (Adapted from Vargas *et al.*, 1982.)

during pregnancy was maintained at levels comparable to those found at sea level. Since hyperventilation is exhibited by European females with a short history of residence at high altitude, it is not likely to have a genetic basis. Instead, it is most likely to reflect primarily the innate biological plasticity and adaptive capacity of females.

Pulmonary diffusing capacity

It is technically difficult to measure the capacity of oxygen to diffuse from the alveoli into alveolar blood (DL_{O_2}) and therefore pulmonary diffusing capacity is frequently assessed indirectly through measurement of the diffusing capacity of carbon monoxide (DL_{CO}) The first assessment of pulmonary diffusing capacity in highlanders was made by Velasquez (1956), who found that DL_{O_2} during submaximal exercise was substantially larger in highland than in lowland Andean Indian adults. Later studies of DL_{CO} also found significantly higher rates of diffusion among highland Andean, Caucasian and Ladakhi adult males (for example, Remmers & Mithoeffer, 1969; DeGraff *et al.*, 1970; Guleria *et al.*, 1971; Vincent *et al.*, 1978).

As noted earlier, highland children exhibit enlarged lungs and lung volumes at an early age. Vargas and colleagues (1982) demonstrated that the early proliferation of alveoli and growth in lung size in Andean highlanders is associated with an increase in pulmonary diffusing capacity. In particular, DL_{CO} was assessed in 125 highland Andean children of both sexes and between 4 and 6 years of age. As indicated in Fig. 5.3, both male and female highland children have substantially larger pulmonary diffusing capacities than lowlanders of the same age and size.

Given that diffusing capacity is larger in males than females at low altitudes, a surprising finding of the Vargas et al. (1982) study was that DL_{CO} was significantly larger in females than in males (Fig. 5.3). Similar sex differences in DL_{CO} have been found in adults (Spielvogel et al., 1977). One possible explanation for these unexpected sex differences is that, due to the increased oxygen demands of pregnancy and the continuous need to maintain adequate levels of oxygen flow to the foetus, there has been selection for highland Andean females with more efficient oxygen transport systems (Haas, 1980).

Oxyhaemoglobin dissociation curve

Oxygen saturation (S_{O_2}) refers to the amount of oxygen attached to haemoglobin relative to the maximal capacity of haemoglobin to bind oxygen, expressed as a percentage. Arterial S_{O_2} at sea level is about 97 % but as altitude increases and alveolar P_{O_2} decreases, there is a corresponding decrease in arterial S_{O_2}. The oxyhaemoglobin dissociation (or association) curve (ODC) describes the relationship between P_{O_2} and S_{O_2} under specified physiological conditions and is therefore a useful way to summarize the oxygen binding and unloading capacities of haemoglobin. Rather than construct the entire curve, researchers sometimes measure one point on the curve, or the P_{O_2} at which haemoglobin is 50 % saturated with oxygen (P_{50}). If physiological conditions are such that there is an increase in S_{O_2} for a given P_{O_2}, this would be reflected in a left-shift of the ODC, and therefore in an increase in the amount of oxygen loaded on to haemoglobin at the alveolar–arterial interface. On the other hand, if physiological conditions are such that there is a decrease in S_{O_2} for a given P_{O_2}, this would be reflected in a right-shift of the ODC, and therefore in an increase in the amount of oxygen unloaded from haemoglobin at the arterial–tissue interface.

Early studies in the Andes indicated that the ODC of native highlanders was shifted to the right, which was interpreted as an adaptive response to hypobaric hypoxia (Aste-Salazar & Hurtado, 1944; Lenfant et al., 1968; Torrance et al., 1970/71). However, Winslow et al. (1981) suggested that these studies had not adequately controlled for the in vivo physiological conditions of highlanders. The enzyme 2,3-diphosphoglycerate (2,3-DPG), which acts to facilitate the unloading of oxygen from haemoglobin at the tissue level, is present in increased amounts in highlanders (Lenfant et al., 1968). Since it facilitates oxygen unloading at the tissue level, 2,3-DPG has the effect of shifting the ODC to the right. Winslow and colleagues argued that previous researchers had controlled for the increase in 2,3-DPG but did not consider the counteracting effect of the slight alkalosis found in highlanders. After adjusting for both 2,3-DPG and pH, Winslow and colleagues (1981) found that P_{50} in highlanders (30.1) was still larger than in lowlanders (29.2), indicating a slight right shift in the ODC, but the difference was much smaller than found by previous researchers and was statistically insignificant. However, Beard, Hurtado, Gomez and Haas (1986) demonstrated that one of the responses of Andean highlanders to a decrease in haemoglobin production as a result of protein-energy malnutrition was a right-shift in the ODC. It is therefore possible that the contradictory findings regarding the effect of hypobaric hypoxia on the ODC could be due to differences between samples in nutritional status.

Further research will be needed to resolve this issue and, in particular, to identify the agents which might modify the effect of hypoxia. Whether it is concluded that hypobaric hypoxia results in no change or a right shift in the ODC, it is worth noting that the most adaptive response might be a left-shifted ODC. This is the pattern observed in native camelids (Banchero & Grover, 1972). In addition, individuals with a genetically determined left-shifted ODC (and therefore a high affinity for oxygen) exhibited a highly adaptive response to chronic hypoxia, as indicated by minimal changes in S_{O_2} and $\dot{V}_{O_2}$max (Hebbel *et al.*, 1978).

Oxygen transport capacity

Oxygen transport capacity has most frequently been assessed through measurement of haemoglobin concentration (g/dl). Hurtado, Merino & Delgado (1945) demonstrated that there is a strong positive curvilinear relationship between altitude and haemoglobin concentration in Andean natives. For example, mean haemoglobin concentration was found to be about 29 % higher than at sea level in highlanders residing in Morochoca, Peru (4530 m). Hurtado recognized that an increase in haemoglobin concentration results in an increase in blood viscosity, and that there are therefore functional limitations to the possible magnitude of adaptive increases in haemoglobin concentration. However, he believed that the enhanced haemoglobin concentration found in most Andean highlanders was not of sufficient magnitude to hinder blood movement and instead increased oxygen transport.

Later studies in the Andes and the Tien Shan mountains also found a positive relationship between altitude and haemoglobin concentration (for example, Garruto, 1976; Mirrakhimov, 1978; Arnaud, Quilici & Riviere, 1981; Garruto & Dutt, 1983; Tufts *et al.*, 1985). However, not all researchers have found such large increases in haemoglobin concentration among Andean highlanders as those found by Hurtado. For example, Garruto (1976) found that haemoglobin concentrations among Nuñoa agriculturalists were only 10–12 % higher than at sea level. Garruto & Dutt (1983) argued that mining communities (such as Morococha) are characterized by a mixed genetic heritage, an increased proportion of relatively recent lowland migrants and an increased frequency of respiratory disorders. They recognize that Hurtado attempted to exclude individuals with respiratory disease or a past history of working in the mines but feel that his sample was nevertheless biased and not representative of most Andean highlanders. Support for this hypothesis was recently provided by Frisancho (1988), who demonstrated that haemoglobin concentration was significantly higher in mining communities than in non-mining communities in the Andes.

Although there is some disagreement about the magnitude of the increase in haemoglobin concentration in Andean highlanders, there is no question that Andean highlanders living under conditions of chronic hypoxia have significantly higher haemoglobin concentrations than found at sea level. However, the results of a number of studies suggest that Himalayan highlanders may not exhibit as large an increase in haemoglobin concentration as found in the Andes (Beall & Reichsman, 1984). For example, Tibetan pastoral nomads living at 4850–5450 m, or about 1000 m higher than Nuñoa, were found to have haemoglobin concentrations which are less than 1 g/dl higher than found at Nuñoa, or much less than would be predicted on the basis of the Andean data (Beall, Goldstein & Tibetan Academy of Social Sciences, 1987). On the other hand, Frisancho (1988) compiled the available data on mean haemoglobin concentration from studies in both the Andes (excluding studies of individuals from mining communities) and the Himalayas. He then regressed mean haemoglobin concentration against atmospheric P_{O_2} in each area and concluded that these regressions were similar. This suggests that Andean and Himalayan natives living under similar levels of hypobaric hypoxic stress exhibit similar mean haemoglobin concentrations, although this conclusion would have been more compelling if the results of a statistical comparison of these regression lines had been presented.

As demonstrated in a number of studies, the level of haemoglobin concentration has important functional implications at high altitude. For example, Tufts *et al.* (1985) and Beard *et al.* (1988) demonstrated that highland Aymara anaemics have significantly reduced submaximal and maximal work capacities. Another area of research which demonstrates the functional significance of haemoglobin concentration involves the maternal–foetal dyad. As noted earlier, pregnancy is a potentially stressful period for highland females (Haas, 1980). In fact, Ballew & Haas (1986) demonstrated that the haemoglobin concentration in the cord blood of newborns was higher at high altitude than at low altitude, suggesting greater hypoxic stress within the uterine environment at high altitude. In addition, they found higher haemoglobin concentrations in European than Indian newborns, suggesting genetic differences in the buffering capacity of the uterine environment. Also, Moore and colleagues (1982b) found that the mothers of high birthweight European babies in Leadville maintained relatively constant haemoglobin concentrations throughout pregnancy, while the mothers of low birthweight babies exhibited a decrease in haemoglobin concentration during pregnancy, possibly resulting in a decrease in oxygen delivery to the foetus. Finally, Haas (1980) found that maternal haemoglobin concentration was a more important determinant of birthweight at high altitude than at low altitude in Bolivia. In addition, he demonstrated that there were important ethnic differences. First, despite their lower socioeconomic status and smaller body size, highland Aymara women residing in La Paz had larger babies than highland European women residing in La Paz, suggesting less hypoxic stress within the uterine environment of Aymara women. Second, maternal haemoglobin concentration was unrelated to birthweight in highland Aymara females but there was a significant negative relationship between these variables in highland European females.

Oxygen consumption

The functional capacity of the entire oxygen transport system can be evaluated by measuring oxygen consumption during maximal exercise, or $\dot{V}_{O_2}$max (ml/kg/min). A failure to respond successfully to hypoxia at any stage of the oxygen transport system will result in a decrease in $\dot{V}_{O_2}$ max. Other useful measures made during an exercise test include ventilatory equivalent ($\dot{V}_E/\dot{V}_{O_2}$), which assesses the efficiency of ventilation and oxygen pulse ($\dot{V}_{O_2}$/heart rate), which is an index of oxygen transport efficiency.

Maximal exercise tests conducted by Institute of Andean Biology researchers indicated that lowlanders experienced a significant decline in $\dot{V}_{O_2}$max upon ascent to high altitudes. Andean highlanders, however, were found to have $\dot{V}_{O_2}$max values at high altitude which were similar to

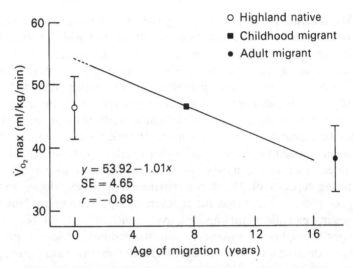

Fig. 5.4. Mean $\dot{V}_{O_2}$max in male adult highland natives and in men who migrated to high altitude as children or as adults. The bars around the mean $\dot{V}_{O_2}$max values of highland natives and adult migrants represent one standard deviation. The line running through the mean $\dot{V}_{O_2}$max of childhood migrants describes the regression relationship between $\dot{V}_{O_2}$max and their age at the time of their migration to high altitude. (Adapted from Frisancho *et al.*, 1973*b*.)

those of lowlanders at low altitude. Studies conducted by IBP researchers (reviewed by Baker, 1976, and Buskirk, 1976) confirmed these findings. In other words, these studies indicated that highlanders had successfully adapted to their hypoxic environment. Based on a comparison of highland natives and migrants, Mazess (1969) suggested that this adaptedness was obtained through developmental adaptation. Frisancho *et al.* (1973*b*) tested this hypothesis by measuring $\dot{V}_{O_2}$max in male high-altitude natives, men who migrated to high altitude during childhood, and men who migrated to high altitude during adulthood. They found that mean $\dot{V}_{O_2}$max was similar between the high-altitude natives and the childhood migrants and that both groups had significantly larger mean $\dot{V}_{O_2}$max values than adult migrants (Fig. 5.4). In addition, there was a significant negative relationship between $\dot{V}_{O_2}$max and age of migration to high altitude in childhood migrants. These findings are consistent with the expectations of developmental adaptation. However, Baker (1976) noted inconsistencies in the results which he suggested might reflect sample bias. In particular, one would predict the $\dot{V}_{O_2}$max of a migrant with an age of migration of 0 years to be similar to the mean $\dot{V}_{O_2}$max of highland natives, but it is in fact substantially larger (Fig. 5.4).

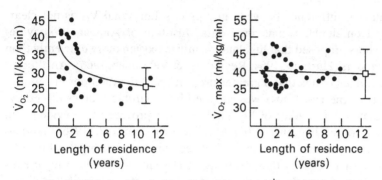

Fig. 5.5. Relationship of submaximal and maximal $\dot{V}_{O_2}$ with length of residence at high altitude in European low-to-high altitude migrant children. (Adapted from Greksa & Haas, 1982, 1983; Haas *et al.*, 1983.)

Several studies were conducted in La Paz, Bolivia, in order to test the developmental adaptation hypothesis further with respect to work capacity. The first study measured submaximal and maximal work capacity in young boys (8.8–13.1 years of age) of European ancestry (Greksa & Haas, 1982, 1983; Haas *et al.*, 1983). The sample included both high-altitude born (HAB) boys and low-altitude born (LAB) boys who later migrated to La Paz. Based on the developmental adaptation hypothesis, one would predict higher $\dot{V}_{O_2}$max values in HAB than in LAB boys. However, after controlling for body size, $\dot{V}_{O_2}$max, as well as measures of ventilation and oxygen transport, did not differ significantly between these groups.

Mean $\dot{V}_{O_2}$max was about 40 ml/kg/min in both HAB and LAB boys. Similar values were later reported by Fellmann *et al.* (1986, 1988) for 10- to 13-year-old highland boys of European ancestry. Mean $\dot{V}_{O_2}$max in lowlanders of the same age varies from about 45 to 50 ml/kg/min (Shephard, 1978). Thus, if developmental adaptation is operating, these data suggest that the process is not completed by an age of about 13 years. In addition, regression analyses indicated that the physiological measures of maximal work performance were not related to length of exposure to hypobaric hypoxia in the LAB boys, as illustrated for $\dot{V}_{O_2}$max in Fig. 5.5. This finding is not consistent with the expectations of developmental adaptation.

Given the general lack of differences between HAB and LAB boys during maximal work, the finding that HAB boys consumed significantly less oxygen than LAB boys when performing the same submaximal work was surprising. Lower oxygen requirements for work within a hypobaric hypoxic environment would clearly be advantageous. The cause of the significant difference between groups in submaximal $\dot{V}_{O_2}$ is not clear. Based on similar submaximal data, Institute of Andean Biology researchers proposed that highlanders utilize oxygen more efficiently than lowlanders (Hurtado, 1964; Reynafarje & Velasquez, 1966). An alternative explanation is that a greater proportion of the total power requirements of the HAB boys was supplied by anaerobic sources of energy. Several recent studies of highland children provide some evidence in support of this hypothesis. In particular, Fellmann *et al.* (1986) found no difference in anaerobic capacity between lowland and highland boys of European ancestry with a mean age of 11 years. However, comparisons of 12-year-olds performing submaximal work detected significant differences between these groups (Fellmann *et al.*, 1988). Mean blood lactate concentration was 35 % higher in highland than in lowland boys and, even more important, the time required to remove one-half of the exercise-induced lactic acid was about 64 % longer in lowland than in highland boys. In addition, anaerobic capacity within the sample of highland boys was positively related to stage of maturation, suggesting that it is acquired developmentally.

In addition to differing significantly between LAB and HAB boys, $\dot{V}_{O_2}$ was also substantially more variable in LAB than in HAB boys. One possible cause of the increased variation in LAB boys is their differing lengths of exposure to hypobaric hypoxia. Although no significant relationships were found when analysing maximal work performance, submaximal $\dot{V}_{O_2}$ (Fig. 5.5), $\dot{V}_E$, ventilation equivalent, and respiratory rate were all significantly related to length of residence at high altitude in the LAB boys, such that increased exposure was associated with improved performance (Greksa & Haas, 1983). All relationships except for that with ventilation equivalent were curvilinear, suggesting a critical length of exposure is involved (Baker, 1976). Examination of the curvilinear relationships suggested that about 5 years of exposure to hypobaric hypoxia are required before the submaximal work performance of the average migrant is equivalent to that of the average HAB boy. These data thus suggest that some of the variation in level of adaptedness to hypobaric hypoxia of the LAB sample, as measured by submaximal exercise performance, is a function of varying lengths of residence at high altitude. However, the proportion of the variance in submaximal work performance which was explained by length of residence at high altitude varied from small to moderate (18–41 %), indicating that factors other than length of residence were also operating.

The submaximal results of this study were generally consistent with the expectations of developmental adaptation while the maximal exercise

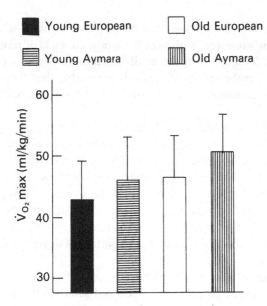

Fig. 5.6. Mean $\dot{V}_{O_2}$max (±one standard deviation) in European and Aymara highland youths. (Adapted from Greksa *et al.*, 1985.)

responses were not. One possible explanation for these contradictory results is that any developmental responses to hypoxia that affect maximal oxygen consumption occur during adolescence, or at an older age than that of the boys tested in this study. In order to test this possibility, a second study was performed in which the sample consisted of both young (11- to 13-year-old) and old (17- to 20-year-old) boys of both European and Aymara ancestry (Greksa, Spielvogel & Paredes-Fernandez, 1985). This research design permitted the simultaneous testing for both developmental and genetic effects.

As noted earlier, mean $\dot{V}_{O_2}$max in untrained lowland boys varies from about 45 to 50 ml/kg/min (Shephard, 1978). Mean $\dot{V}_{O_2}$max in younger and older Aymara boys, as well as older European boys, were within this range (Fig. 5.6). Younger Europeans, however, fell below the lowland norm (Fig. 5.6), indicating incomplete adaptation in this group, which is consistent with the findings of the previous study.

Mean $\dot{V}_{O_2}$max was greater in Aymara than European boys of the same age but the differences were not statistically significant, suggesting that the relatively high $\dot{V}_{O_2}$max values of Aymara highlanders do not have a genetic basis. However, there were statistically significant differences between ethnic groups in the functional capacity of specific components of the oxygen transport system. In particular $\dot{V}_E$, oxygen pulse and ventilation equivalent were all significantly higher in Aymara than in European boys of the same age. Although these physiological differences do not result in differences in the overall functioning of the oxygen transport system, as evaluated by $\dot{V}_{O_2}max$, it is possible that they may have functional consequences at a later stage of life (Greksa & Beall, 1989).

$\dot{V}_{O_2}max$, $\dot{V}_E$ and ventilation equivalent were significantly greater in older than younger boys in both ethnic groups, which is consistent with the expectations of developmental adaptation. However, a much stronger argument for developmental adaptation could be made if it could be demonstrated that the magnitude of the increase in $\dot{V}_{O_2}max$ between early and late adolescence in highlanders was significantly greater than is found at low altitude. Unfortunately, that comparison cannot be made due to uncertainties about the pattern of development of $\dot{V}_{O_2}max$ at low altitude.

Finally, young adult highland males have $\dot{V}_{O_2}max$ values similar to those of lowlanders, suggesting they have adapted successfully to hypobaric hypoxia. However, since $\dot{V}_{O_2}max$ decreases during adulthood, a final important question is whether this level of adaptedness is maintained throughout adulthood. This really involves two related questions. First, is the rate of decrease in $\dot{V}_{O_2}max$ during adulthood at high altitude similar to that found at low altitude? Second, given that $\dot{V}_{O_2}max$ (and therefore the overall functional capacity of the oxygen transport system) decreases during adulthood, are older adults as well adapted to hypobaric hypoxia as young adults? The latter question cannot yet be answered, although there is some suggestion that aging might be associated with increased hypoxic stress in highlanders (Greksa & Beall, 1989). However, although few studies have been conducted, the rate of decrease in $\dot{V}_{O_2}max$ during adulthood at high altitude appears to be similar to that found at low altitudes (Lange-Anderson, 1972; Weitz, 1973; Greksa *et al.*, 1984).

Conclusion

All of the available data indicates that adult highland natives are less stressed by, and therefore better adapted to, high-altitude environments than are low-to-high altitude migrants. An important theoretical issue for the discipline of human biology involves determining the process by which this superior level of adaptation is obtained. The primary candidate for a number of years has been developmental adaptation, as proposed and primarily developed by Roberto Frisancho (1976). A conclusive evaluation of this hypothesis will require longitudinal studies, especially of migrants between altitude zones. Nevertheless, in the opinion of the author, the available data are generally consistent with the developmental adaptation hypothesis.

Assessments of the developmental adaptation hypothesis would ideally be based on *in vivo* monitoring of oxygen utilization. Only developmental changes which enhanced oxygen utilization would then be considered adaptive. Unfortunately, this is not a technically feasible approach. However, it is possible to measure $\dot{V}_{O_2}$max, which indirectly assesses the functional capacity of the overall oxygen transport system. It is important to emphasize that arguments about high-altitude adaptation should ultimately focus on such integrative measures. Changes in individual components of the oxygen transport system may well be involved in the adaptive process but such changes should not be used as evidence of adaptation in and of themselves. The functional capacity of any system, including the oxygen transport system, is the result of continuous and complex interactions between all of its components. As a result, a logically beneficial change in any one component of the oxygen transport system could be counteracted by changes in other components. This is one possible explanation for the failure consistently to detect evidence for developmental adaptation in studies of lung volumes at high altitude.

Thus, prior to evaluating the relative importance of developmental changes in specific components of the oxygen transport system, it is necessary to demonstrate that developmental responses result in enhancements of the functional capacity of the oxygen transport system as a whole. Although it has limitations, $\dot{V}_{O_2}$max is the best available index of the overall functional capacity of the oxygen transport system. There is strong evidence that young highland adult males have $\dot{V}_{O_2}$max values similar to those of lowlanders at low altitude (Baker, 1976). (The same is probably true of females but, unfortunately, due to the paucity of information on females of any age, this must remain an assumption.) On the other hand, several studies of both Aymara and European children between the ages of 9 and 13 years (that is, late childhood to early adolescence) suggest that children of this age have $\dot{V}_{O_2}$max values lower than generally found in lowland children of the same age (Greksa & Haas, 1982; Greksa *et al.*, 1985; Fellmann *et al.*, 1986, 1988). In other words, it appears that highlanders may develop $\dot{V}_{O_2}$max values similar to those of lowlanders during adolescence. If true, this conclusion is, of course, consistent with the developmental adaptation hypothesis as proposed by Frisancho. However, even if one assumes that $\dot{V}_{O_2}$max increases to lowland values during adolescence, this does not necessarily mean that all of the developmental responses which led to this enhancement of $\dot{V}_{O_2}$max also occurred during adolescence. Instead, it seems likely that different components of the oxygen transport system are affected at different stages of development. For example, lung volumes and pulmonary diffusing capacity appear to be enhanced at very early ages (Vargas *et al.*, 1982), while a blunted hypoxic drive and an enhanced anaerobic capacity are not developed until adolescence (Lahiri *et al.*, 1976; Fellmann *et al.*, 1988).

It is worth noting that, if this interpretation is correct, the period of life from conception through early childhood may be a highly stressful one for highlanders. As indicated in the preceding review, some physiological research has been conducted on this period of growth and development but additional research may be enlightening.

A second important theoretical issue for the field of human biology involves determining whether there are population differences, particularly between Andean and Himalayan natives, in the nature of their adaptation to hypobaric hypoxia. There is less $\dot{V}_{O_2}$max data available for the Himalayas than for the Andes but, based on this measure of the functional capacity of the oxygen transport system, both groups appear to be equally well adapted to their hypobaric hypoxic environments. However, there is some evidence that there are differences between groups in the nature of their responses to hypobaric hypoxia. In particular, there may be differences between Himalayan and Andean natives in lung volumes, haemoglobin concentration and ventilatory response to hypoxia. Further research is needed to confirm these apparent differences but, given the complexity of the oxygen transport system, it is not really surprising that multiple adaptive pathways might exist. Expressed differently, there is no reason to assume that there is a single, rigidly defined adaptive pattern which is characteristic of all high-altitude populations.

Finally, it is worth noting one of the practical implications of the unique physiological characteristics of highlanders. The implementation and evaluation of public health programmes often requires the identification of those at risk and therefore most likely to benefit from intervention. This is frequently accomplished by identifying those individuals who fall below a specific cut-off value derived from appropriate reference standards. However, due to the physiological consequences of lifelong residence at high altitudes, cut-off values based on low-altitude reference standards are not always appropriate. For example, measures of pulmonary function based on lung volumes can be used as a screening device for the identification of obstructive lung disease. However, since lung volumes are significantly larger at high than at low altitudes (Frisancho, 1976), cut-off values based on lowland populations are not appropriate for high-altitude populations (Kryger *et al.*, 1978). A second example involves the use of haemoglobin concentration to identify anaemic individuals (Haas *et al.*, 1988). Since there is a positive curvilinear relationship between altitude and haemoglobin concentration in the Andes (Hurtado *et al.*, 1945), cut-off values for anaemia based on low-altitude reference standards are inappropriate for highland populations.

Although altitude-specific lung function and haemoglobin concentration reference standards are clearly needed, establishing those standards is no simple matter. There are several problems. First, as noted earlier, the magnitude of the effect of hypobaric hypoxia on lung volumes and haemoglobin concentration is far from clear. Thus, estimates of the prevalence of individuals at risk can vary greatly, depending on how the reference standard is defined. For example, Haas *et al.* (1988) found prevalences of anaemia for a La Paz population which varied from 0.0 % to 23.4 %, depending on which of several different suggested high-altitude cut-off values were utilized. Second, there appear to be ethnic differences between high-altitude populations in both lung volumes (Greksa & Beall, 1989) and haemoglobin concentration (Beall *et al.*, 1987; Haas, 1980). Once again, however, the magnitude of inter-population differences is not clear. Finally, the pattern of change with age in some physiological parameters may be different at high altitudes than found at low altitudes (Beall, 1984), thus requiring further modification of reference standards. Thus, at least for lung volumes and haemoglobin concentration (and probably for other measures with significance for public health programmes), reference standards which are altitude-, population- and perhaps age-specific may be required. However, additional research on high-altitude physiology will be required before it will be possible to construct such reference standards.

References
Arnaud, J., Quilici, J. C. & Riviere, G. (1981). High-altitude haematology: Quechua-Aymara comparisons. *Annals of Human Biology*, **8**, 573–8.
Aste-Salazar, H. & Hurtado, A. (1944). The affinity of hemoglobin for oxygen at sea level and at high altitudes. *American Journal of Physiology*, **142**, 733–44.
Baker, P. T. (1969). Human adaptation to high altitude. *Science*, **163**, 1149–56.
  (1976). Work performance of highland natives. In *Man in the Andes: a multidisciplinary study of high-altitude Quechua*, ed. P. T. Baker & M. A. Little, pp. 300–14. Stroudsburg: Dowden, Hutchinson & Ross.

(1978). The Biology of High-altitude Peoples. Cambridge: Cambridge University Press.

(1988). Human adaptability. In Human Biology (3rd edn), ed. G. A. Harrison, J. M. Tanner, D. R. Pilbeam & P. T. Baker, pp. 439–547. Oxford: Oxford University Press.

Ballew, C. & Haas, J. D. (1986). Hematologic evidence of fetal hypoxia among newborn infants at high altitude in Bolivia. American Journal of Obstetrics and Gynecology, 155, 166–9.

Banchero, N. & Grover, R. F. (1972). Effects of different levels of simulated altitude on oxygen transport in llama and sheep. American Journal of Physiology, 222, 1239–45.

Beall, C. M. (1984). Aging and growth at high altitudes in the Himalayas. In The People of South Asia, ed. J. R. Lukacs, pp. 365–85. New York: Plenum Press.

Beall, C. M., Goldstein, M. C. & the Tibetan Acadamy of Social Sciences (1987). Hemoglobin concentration of pastoral nomads permanently resident at 4850–5450 m in Tibet. American Journal of Physical Anthropology, 73, 433–8.

Beall, C. M. & Reichsman, A. B. (1984). Hemoglobin levels in a Himalayan high altitude population. American Journal of Physical Anthropology, 63, 301–6.

Beall, C. M., Strohl, K. P. & Brittenham, G. M. (1983). Reappraisal of Andean high altitude erythrocytosis from a Himalayan perspective. Seminars in Respiratory Medicine, 5, 195–201.

Beard, J. L., Haas, J. D., Tufts, D., Spielvogel, H., Vargas, E. & Rodriquez, C. (1988). Iron deficiency anemia and steady-state work performance at high altitude. Journal of Applied Physiology, 64, 1878–84.

Beard, J. L., Hurtado, A., Gomez, L. & Haas, J. D. (1986). Functional anemia of complicated protein-energy malnutrition at high altitude. American Journal of Clinical Nutrition, 44, 181–7.

Boyce, A. J., Haight, J. S. J., Rimmer, D. B. & Harrison, G. A. (1974). Respiratory function in Peruvian Quechua Indians. Annals of Human Biology, 1, 137–48.

Buskirk, E. R. (1976). Work performance of newcomers to the Peruvian highlands. In Man in the Andes: a multidisciplinary study of high-altitude Quechua, ed. P. T. Baker & M. A. Little, pp. 283–99. Stroudsburg: Dowden, Hutchinson & Ross.

Cerretelli, P. (1980). Gas exchange at high altitude. In Pulmonary Gas Exchange, ed. J. B. West, pp. 97–145. New York: Academic Press.

Clegg, E. J., Pawson, I. G., Ashton, E. H. & Flinn, R. M. (1972). The growth of children at different altitudes in Ethiopia. Philosophical Transactions of the Royal Society (London), 264, 403–37.

Cudkowicz, L., Spielvogel, H. & Zubieta, G. (1972). Respiratory studies in women at high altitude (3600 m or 12 200 ft and 5200 m or 17 200 ft). Respiration, 28, 393–426.

DeGraff, A. C., Grover, R. F., Johnson, R. L., Hammon, J. W. & Miller, J. M. (1970). Diffusing capacity of the lung in Caucasians native to 3100 m. Journal of Applied Physiology, 29, 71–6.

Fellmann, N., Bedu, M., Spielvogel, H., Falgairette, G., Van Praagh, E. & Coudert, J. (1986). Oxygen debt in submaximal and supramaximal exercise in children at high and low altitude. *Journal of Applied Physiology*, **60**, 209–15.

Fellmann, N., Bedu, M., Spielvogel, H., Falgairette, G., Van Praagh, E., Jarrige J. F. & Coudert, J. (1988). Anerobic metabolism during pubertal development at high altitude. *Journal of Applied Physiology*, **64**, 1382–6.

Frisancho, A. R. (1969). Human growth and pulmonary function of a high altitude Peruvian Quechua population. *Human Biology*, **41**, 365–79.

(1976). Growth and morphology at high altitude. In *Man in the Andes: a multidisciplinary study of high-altitude Quechua*, ed. P. T. Baker and M. A. Little, pp. 180–207. Stroudsburg: Dowden, Hutchinson & Ross.

(1988). Origins of differences in hemoglobin concentrations between Himalayan and Andean populations. *Respiration Physiology*, **72**, 13–15.

Frisancho, A. R. & Baker, P. T. (1970). Altitude and growth: a study of the patterns of physical growth of a high-altitude Peruvian Quechua population. *American Journal of Physical Anthropology*, **32**, 279–92.

Frisancho, A. R., Martinez, C., Velasquez, T., Sanchez, J. & Montoye, H. (1973b). Influence of developmental adaptation on aerobic capacity at high altitude. *Journal of Applied Physiology*, **34**, 176–80.

Frisancho, A. R., Velasquez, T. & Sanchez, J. (1973a). Influence of developmental adaptation on lung function at high altitude. *Human Biology*, **45**, 583–94.

Garruto, R. M. (1976). Hematology. In *Man in the Andes: a multidisciplinary study of high-altitude Quechua*, ed. P. T. Baker and M. A. Little, pp. 261–82. Stroudsburg: Dowden, Hutchinson & Ross.

Garruto, R. M. & Dutt, J. S. (1983). Lack of prominent compensatory polycythemia in traditional native Andeans living at 4200 meters. *American Journal of Physical Anthropology*, **61**, 355–66.

Greksa, L. P. (1988). Effect of altitude on the stature, chest depth and forced vital capacity of low-to-high altitude migrant children of European ancestry. *Human Biology*, **60**, 23–32.

Greksa, L. P. & Beall, C. M. (1989). Development of chest size and lung function at high altitude. In *Human Population Biology: a transdisciplinary science*, ed. M. A. Little & J. D. Haas, pp. 222–38. New York: Oxford University Press.

Greksa, L. P. & Haas, J. D. (1982). Physical growth and maximal work capacity in preadolescent boys at high-altitude. *Human Biology*, **54**, 677–95.

(1983). Work capacity of European boys at altitude. *American Journal of Physical Anthropology*, **60**, 201–2.

Greksa, L. P., Haas, J. D., Leatherman, T. L., Spielvogel, H. & Thomas, R. B. (1984). Work performance of high-altitude Aymara males. *Annals of Human Biology*, **11**, 227–33.

Greksa, L. P., Spielvogel, H., Caceres, E. & Paredes-Fernandez, L. (1987). Lung function of young Aymara highlanders. *Annals of Human Biology*, **14**, 533–42.

140   L. P. Greksa

Greksa, L. P., Spielvogel, H. & Paredes-Fernandez, L. (1985). Maximal exercise capacity in adolescent European and Amerindian high-altitude natives. *American Journal of Physical Anthropology*, 67, 209–16.

Greksa, L. P., Spielvogel, H., Paz-Zamora, M., Caceres, E. & Paredes-Fernandez, L. (1988). Effect of altitude on the lung function of high altitude residents of European ancestry. *American Journal of Physical Anthropology*, 75, 77–85.

Guleria, J. S., Pande, J. N., Sethi, P. K. & Roy, S. B. (1971). Pulmonary diffusing capacity at high altitude. *Journal of Applied Physiology*, 31, 536–43.

Haas, J. D. (1980). Maternal adaptation and fetal growth at high altitude in Bolivia. In *Social and Biological Predictors of Nutritional Status, Physical Growth and Neurological Development*, ed. L. S. Green, pp. 257–89. New York: Academic Press.

Haas, J. D., Greksa, L. P., Leatherman, T. L., Spielvogel, H., Paredes-Fernandez, L., Moreno-Black, G. & Paz-Zamora, M. (1983). Submaximal work performance of native and migrant preadolescent boys at high altitude. *Human Biology*, 55, 517–27.

Haas, J. D., Tufts, D. A., Beard, J. L., Roach, R. C. & Spielvogel, H. (1988). Defining anaemia and its effect on physical work capacity at high altitudes in the Bolivian Andes. In *Capacity for Work in the Tropics*, ed. K. J. Collins & D. F. Roberts, pp. 85–105. New York: Cambridge University Press.

Hackett, P. H., Reeves, J. T., Reeves, C. D., Grover, R. F. & Rennie, D. (1980). Control of breathing in Sherpas at low and high altitude. *Journal of Applied Physiology*, 49, 374–9.

Harrison, G. A., Küchemann, C. F., Moore, M. A. S., Boyce, A. J., Baju, T., Mourant, A. E., Godber, M. J., Glasgow, B. G., Kopec, A. C., Tills, D. & Clegg, E. J. (1969). The effects of altitudinal variation in Ethiopian populations. *Philosophical Transactions of the Royal Society (London)*, 256, 147–82.

Hebbel, R. P., Eaton, J. W., Kronenberg, R. S., Zanjani, E. D., Moore, L. G. & Berger, E. M. (1978). Human llamas: adaptation to altitude in subjects with high hemoglobin oxygen affinity. *Journal of Clinical Investigation*, 62, 593–600.

Hoff, C. (1974). Altitudinal variations in the physical growth and development of Peruvian Quechua. *Homo*, 24, 87–99.

Hsu, K. H. K., Jenkins, D. E., Hsi, B. P., Bourhofer, E., Thompson, V., Tanakawa, N. & Hsieh, G. S. J. (1979). Ventilatory functions of normal children and young adults – Mexican-American, white, and black. I. Spirometry. *Journal of Pediatrics*, 95, 14–23.

Hurtado, A. (1932). Respiratory adaptation in the Indian natives of the Peruvian Andes. Studies at high altitude. *American Journal of Physical Anthropology*, 17, 137–65.

(1964). Animals in high altitudes: resident man. In *Handbook of Physiology, Section 4: Adaptation to the Environment*, ed. D. B. Dill, E. F. Adolph & C. G. Wilber, pp. 843–60. Washington, DC: American Physiological Society.

Hurtado, A., Merino, C. & Delgado, E. (1945). Influence of anoxemia on the hemopoietic activity. *Archives of International Medicine*, 75, 284–323.

Kryger, M., Aldrich, F., Reeves, J. T. & Grover, R. F. (1978). Diagnosis of airflow obstruction at high altitude. *American Review of Respiratory Disease*, **117**, 1055–8.

Lahiri, S. (1984). Respiratory control in Andean and Himalayan high-altitude natives. In *High Altitude and Man*, ed. J. B. West & S. Lahiri, pp. 147–62. Bethesda: American Physiological Society.

Lahiri, S., Brody, J. S., Motoyama, E. K. & Velasquez, T. M. (1978). Regulation of breathing in newborns at high altitude. *Journal of Applied Physiology*, **47**, 673–8.

Lahiri, S., Delaney, R. G., Brody, J. S., Simpser, M., Velasquez, T., Motoyama, E. K. & Polgar, C. (1976). Relative roles of environmental and genetic factors in respiratory adaptation to high altitude. *Nature*, **261**, 133–4.

Lange-Anderson, K. (1972). The effect of altitude variation on the physical performance capacity of Ethiopian men. In *Human Biology of Environmental Change*, ed. D. J. M. Vorster, pp. 154–63. London: Unwin Brothers.

Lenfant, C., Torrance, J., English, E., Finch, C. A., Reynafarje, C., Ramos, J. & Faura, J. (1968). Effect of altitude on oxygen binding by hemoglobin and on organic phosphate levels. *Journal of Clinical Investigation*, **47**, 2652–6.

Little, M. A. (1982). The development of ideas about human ecology and adaptation. In *A History of American Physical Anthropology, 1930–1980*, ed. F. Spencer, pp. 405–33. New York: Academic Press.

Mazess, R. B. (1969). Exercise performance at high altitude in Peru. *Federation Proceedings*, **28**, 1301–6.

Mirrakhimov, M. M. (1978). Biological and physiological characteristics of the high-altitude natives of Tien Shan and the Pamirs. In *The Biology of High-altitude Populations*, ed. P. T. Baker, pp. 299–315. Cambridge: Cambridge University Press.

Monge M., C. (1948). *Acclimatization in the Andes*. Baltimore: Johns Hopkins University Press.

Moore, L. G., Brodeur, P., Chumbe, O., D'Brot, J., Hofmeister, S. & Monge, C. (1986). Maternal hypoxic ventilatory response, ventilation, and infant birth weight at 4300 m. *Journal of Applied Physiology*, **60**, 1401–6.

Moore, L. G., Jahnigen, D., Rounds, S. S., Reeves, J. T. & Grover, R. F. (1982*a*). Maternal hyperventilation helps preserve arterial oxygenation during high-altitude pregnancy. *Journal of Applied Physiology*, **52**, 690–4.

Moore, L. G., Rounds, S. S., Jahnigen, D., Grover, R. F. & Reeves, J. T. (1982*b*). Infant birth weight is related to maternal arterial oxygenation at high altitude. *Journal of Applied Physiology*, **52**, 695–9.

Mueller, W. H., Yen, F., Rothhammer, F. & Schull, W. J. (1978). A multinational Andean genetic and health program. VI. Physiological measurements of lung function in an hypoxic environment. *Human Biology*, **50**, 489–513.

Remmers, J. E. & Mithoeffer, J. C. (1969). The carbon monoxide diffusing capacity in permanent residents at high altitudes. *Respiration Physiology*, **6**, 233–44.

Reynafarje, B. & Velasquez, T. (1966). Metabolic and physiological aspects of exercise at high altitude. I. Kinetics of blood lactate, oxygen consumption and oxygen debt during exercise and recovering breathing air. *Federation Proceedings*, **25**, 1397–9.

Shephard, R. J. (1978). *Human Physiological Work Capacity*. Cambridge: Cambridge University Press.

Spielvogel, H. S., Vargas, E., Antezana, G., Barragan, L. & Cudkowicz, L. (1977). Effect of posture on pulmonary diffusing capacity and regional distribution of pulmonary blood flow in normal male and female high altitude dwellers at 3650 m (12 200 ft). *Respiration*, **34**, 125–35.

Sun, S., Zhang, J.-G., Zhoma, Tao, McCullough, R. E., McCullough, R. G., Reeves, C. S., Reeves, J. T. & Moore, L. G. (1988). Higher ventilatory drives in Tibetan than male residents of Lhasa (3658 m). *American Review of Respiratory Disease*, **137**, 410.

Torrance, J. D., Lenfant, C., Cruz, J. & Marticorena, E. (1970/71). Oxygen transport mechanisms in residents at high altitude. *Respiration Physiology*, **11**, 1–15.

Tufts, D. A., Haas, J. D., Beard, J. L. & Spielvogel, H. (1985). Distribution of hemoglobin and functional consequences of anemia in adult males at high altitude. *American Journal of Clinical Nutrition*, **42**, 1–11.

Vargas, E., Beard, J., Haas, J. & Cudkowicz, L. (1982). Pulmonary diffusing capacity in young Andean highland children. *Respiration*, **43**, 330–5.

Velasquez, T. (1956). *Maximal Diffusing Capacity of the Lungs at High Altitudes*. AF SAM Report 56–108. Randolph Field, Texas: Air Force School of Aviation Medicine.

——— (1976). Pulmonary function and oxygen transport. In *Man in the Andes: a multidisciplinary study of high-altitude Quechua*, ed. P. T. Baker & M. A. Little, pp. 237–60. Stroudsburg: Dowden, Hutchinson & Ross.

Vincent, J., Hellot, M. F., Vargas, E., Gautier, H., Pasquis, P. & Lefrancois, R. (1978). Pulmonary gas exchange, diffusing capacity in natives and newcomers at high altitude. *Respiration Physiology*, **34**, 219–31.

Weil, J. V., Bryne-Quinn, E., Sodal, I. E., Filley, G. F. & Grover, R. F. (1971). Acquired attenuation of chemoreceptor function in chronically hypoxic man at high altitude. *Journal of Clinical Investigation*, **50**, 186–95.

Weitz, C. A. (1973). 'The effects of aging and habitual activity pattern on exercise performance among a high-altitude Nepalese population.' PhD Dissertation, State College, Pennsylvania State University.

Winslow, R. M., Monge C., C., Statham, N. J., Gibson, C. F., Charache, S., Whittembury, J., Moran, O. & Berger, R. L. (1981). Variability of oxygen affinity of blood: human subjects native to high altitude. *Journal of Applied Physiology*, **51**, 1411–16.

6 Darwinian fitness, physical fitness and physical activity

ROBERT M. MALINA

Human evolution, and hence natural selection through Darwinian fitness, is sometimes considered the central concept of biological anthropology. As the study of anthropology has turned increasingly to contemporary human population biology, and especially to more practical issues related to the attainment and maintenance of health, interest in the study of variation in different physiological functions has greatly increased. Therefore, studies of populations in a variety of ecological settings and of individuals exposed to unique stresses such as those associated with rigorous training for sport are now a mainstay of contemporary biological anthropology. Such studies, needless to say, also contribute significantly to the development of general theory about human biology and human variability.

The so-called 'healthy life-style', which includes physical activity as an essential component, is an objective of many, especially in developed countries. This is quite apparent in the relatively rapid growth of the 'physical fitness industry', including sales of exercise equipment (especially for the home), weight control centres, comprehensive adult and/or corporate fitness programmes, and so on. Further, current research on risk factors for several degenerative diseases, such as cardiovascular disease, obesity and non-insulin-dependent diabetes, indicates a potentially beneficial role for regular physical activity. Several diseases, especially of the cardiovascular system, may be prevented, or their development delayed, by the incorporation of sound programmes of regular physical activity, among other aspects of life-style such as diet.

Habitual physical activity is related to physical fitness (Blair, 1988). A regular programme of physical activity aimed at conditioning the heart and circulatory system, for example, ordinarily results in an improved level of cardiovascular fitness. Although cardiovascular fitness is only one component of physical fitness (see below), it is perhaps the most important from the perspective of disease prevention. However, opinions differ concerning the relative importance of physical activity or cardiovascular fitness in the risk for cardiovascular disease.

Given the apparent association between physical activity and physical fitness and health and disease, the physically active and fit life-style is often viewed as important to longevity. This would seem to imply a literal interpretation of the 'survival of the fittest'. Further, emphasis on the fit physique and slenderness, which is especially apparent in commercials and advertisements for fitness equipment and nutrition programmes, would seem to suggest a relationship between physical fitness and sexual selection, perhaps even with sexual prowess. Indeed, level of habitual physical activity and physical fitness may be factors in mate selection.

On the other hand, evidence from highly trained female athletes and some recreational runners indicates a relatively high prevalence of secondary amenorrhoea and possibly impaired fecundity. Further, hormonal changes in male marathon runners appear to be quite similar to those that are related to temporary infertility in female athletes. Thus, excessive, prolonged physical training may be associated with altered reproductive function in both sexes, and thus have a potentially negative effect on fertility.

It it thus reasonable to inquire into the association, if any, between physical activity and physical fitness, on one hand, and Darwinian or genetic fitness, on the other hand. The two kinds of fitness are obviously not identical; they may, in fact, be in opposition. The importance of relating physical fitness to Darwinian fitness is not only, and not even primarily, because of confusion between different meanings of the term 'fitness'. On the contrary, the long-term welfare of the human species, which is achieved through fitness to survive and reproduce, is intimately related to the immediate fitness of individuals to respond to the physical, physiological, social, psychological and emotional challenges of everyday life. Thus, several aspects of the possible association between Darwinian fitness and physical fitness are subsequently considered. After briefly defining several concepts related to each type of fitness, the following topics will be considered:

1 assortative mating for physical fitness and activity;
2 training and sexual maturation in boys and girls;
3 training and menstrual dysfunction;
4 training during pregnancy;
5 sex hormones in trained males; and
6 the relationship between physical activity and longevity.

Definitions

Physical fitness

The definition of physical fitness initially offered by Clarke (1971) and slightly modified by the American Academy of Physical Education is one accepted by many in physical education and the sport sciences:

> Physical fitness is the ability to carry out daily tasks with vigor and alertness, without undue fatigue and with ample energy to engage in leisure time pursuits and to meet the above average physical stresses encountered in emergency situations.
> (*Journal of Physical Education and Recreation*, 1979, p. 28.)

The definition is general and perhaps a bit too all-encompassing. More recently, physical fitness has been subdivided into *health-related fitness* and *performance-related fitness*. The former presumably includes components which, to some extent, prevent the development of diseases related to physical inactivity, that is, degenerative diseases of the heart and blood vessels such as coronary heart disease, musculoskeletal disorders including low back pain and limited ranges of movement, and obesity. Thus, health-related fitness is operationalized as including cardiovascular endurance, muscular strength and endurance, flexibility and body composition, specifically fatness (AAHPERD, 1980, 1984). Performance-related fitness includes various components of skilled movement – that is, agility, balance, coordination, power, speed and strength – which permit the individual to perform a wide range of physical activities (AAHPER, 1976; CAHPER, 1980; Simons & Renson, 1982; Reiff *et al.*, 1986). The specific components are ordinarily viewed in the context of performance on specific motor tests, for example, 50-yard dash as a measure of speed, the shuttle run as a measure of agility, or the standing long jump as a measure of power. Performance-related fitness tests are most commonly used with children, while health-related fitness tests are used with both children and adults.

Physical fitness is a dynamic concept. An individual's level of fitness can change. Some change occurs as a function of growth and maturation, and of aging. Diseases such as those indicated above can contribute to reduced levels of physical fitness both directly and indirectly. On the other hand, alterations in life-style, specifically patterns of habitual physical activity and diet, may improve physical fitness and perhaps slow the progress of several degenerative diseases.

Physical activity and training

Physical activity and training are not synonymous. Physical activity refers to 'any bodily movement produced by the skeletal muscles that results in energy expenditure' (Caspersen, Powell & Christenson, 1985). Physical activity thus includes movements associated with work and leisure, in addition to movements occurring during sleep. Physical activities, of course, vary in type: for example, they include household chores, occupational tasks, calisthenics, sports, and so on. They also vary considerably in intensity and duration. Of interest to studies of health and physical fitness is an individual's pattern of habitual physical activity.

Although an individual may participate regularly in physical activity, this may not qualify as training. Training refers to regular, systematic practice of physical activities such as calisthenics, lifting weights, running and games/sports at specific intensities and for specific periods of time, quite often with competition as the objective. Training programmes are generally specific, involving endurance, strength, and sports skill training, but they may include a variety of training stimuli, for example, both strength and endurance training in swimming, or endurance and skill training in basketball. Training is not a single entity; rather, it can be viewed as a continuum from relatively mild to severely stressing work. Programmes need to be specifically defined in terms of intensity, duration and volume.

In most training studies, programmes are generally short-term, lasting several weeks to several months. Rarely is the training stimulus monitored over several years. An exception, of course, is elite athletes who typically train over several years. Outstanding athletes in a number of sports often begin systematic training at an early age. Swimmers and gymnasts, for example, often begin regular training in these sports as early as 6 and 7 years of age. Although studies of athletes provide significant information, it must be emphasized that elite athletes are a highly select group, primarily for skill but also for size and physique in some sports, and thus differ in many respects from the general population.

High levels of habitual physical activity are generally associated with high levels of health- and performance-related fitness, and vice versa. There is, however, a significant genotypic component of physical fitness (Bouchard & Malina, 1983; Bouchard, 1986; Malina, 1986), and significant genotype–environment interactions in twin studies suggest that sensitivity to training is dependent to some extent upon the individual's genotype (Bouchard, 1986).

Darwinian fitness

Darwinian fitness is an evolutionary and a genetic concept. It refers to reproductive efficiency and fertility, and is generally defined in terms of the average reproductive success of the individual. Although something of an oversimplification, Dobzhansky (1962, p. 125) offers a reasonable description of Darwinian fitness:

> . . . the genetic fitness of a genotype, and by extension of an individual, is measured by the contribution it makes, relative to other genotypes or individuals, to the gene pool of the succeeding generations. The fittest is the parent of the greatest number of surviving children.

Thus, increased Darwinian fitness refers to an increased number of fertile offspring in the next generation, while decreased Darwinian fitness refers to a reduced number of fertile offspring in the next generation. The link between generations is genetic, and genes passed from parents to offspring comprise the link.

Variation in Darwinian fitness occurs due to differential survival/ mortality and differential fertility. Characteristics of an individual which influence his/her ability to survive and/or to reproduce can affect his/her genetic fitness. Genes and the environment, in addition to gene-environment interactions, influence an individual's characteristics and thus genetic fitness.

Assortative mating

Positive assortative mating for ability, education, religion and so on is reasonably well established (Garrison, Anderson & Reed, 1968; Harrison, Givson & Hiorns, 1976; Johnson *et al.*, 1976), as is that for physical characteristics (Susanne, 1967; Spuhler, 1968; Roberts, 1977). Husband–wife correlations for body size and related measurements tend to be low in populations of primarily European ancestry. Spouse correlations for several measures of physical fitness are shown in Table 6.1. The correlations tend to be low, and many are in the same general range as those for physical characteristics, although those reported for strength by Wolanski (1973) are rather high. Spouse correlations for tasks requiring accuracy, speed and precision of movement tend to be in the same range as those for aerobic performance and strength (Malina, 1986).

Spouse correlations for estimates of overall energy expenditure (a proxy for habitual physical activity) are 0.24 and 0.26 in French Canadians, but the correlation for pattern of intense physical activity is lower,

Table 6.1. *Spouse correlations for several tests of physical fitness*

Source	n pairs	Test		r
Montoye & Gayle (1978)	27	$\dot{V}_{O_2}$max, measured or estimated, adjusted for age, weight, fatness		0.18
Lortie *et al.* (1982)	119	$\dot{V}_{O_2}$max/kg, estimated		0.33
Lesage *et al.* (1985)	20	$\dot{V}_{O_2}$max/kg, measured		0.21
Bouchard (1986)	276	Physical working capacity (PWC) 150/kg		0.19*
Wolanski (1973)	36–72	Strength:	right grip	0.74
			left grip	0.77
			shoulder	0.53
			back	0.44
		Balance:	beam walk	−0.05
			turning, timed	0.25
			turning, number	0.04
Kovar (1981)	60	Strength:	grip	0.26
			back	0.26
Szopa (1982)	347	Strength:	right grip	0.15
			left grip	0.26
			grip/weight	0.23
			arm	0.17
			arm/weight	0.17
		Motor:	vertical jump	0.35
Devor & Crawford (1984)	53	Strength: dominant grip		0.01†
		Trunk flexibility		−0.09†

* Intraclass correlation based on analysis of variance.
† Correlations based on standardized residuals.

0.13 (Bouchard, personal communication). Husband–wife correlations for reported levels of physical activity among Anglos and Mexican Americans in the San Antonio Heart Study are 0.13 and 0.20, respectively, and show a tendency to decrease as socioeconomic status increases (Malina, unpublished data). These correlations are reasonably similar to those for aerobic power and strength.

The data thus indicate some degree of positive assortative mating for several components of physical fitness and for level of physical activity. Although correlations are generally low, assortative mating, and especially its genetic component, is of significance because it will not only lead to an increase in the frequency of homozygotes for genetic loci that may be relevant, but will also inflate the additive genetic variance for a given characteristic.

Training and sexual maturation

Sexual maturation is mediated primarily by changes in the neuroendocrine system. Gonadotropin releasing hormone (GnRH) is secreted by the hypothalamus and results in the secretion of gonadotropins by the anterior pituitary. The latter in turn initiate the maturation of the gonads and the process of sexual maturation. The initiation of changes in the hypothalamic–pituitary–gonadal axis resides in the central nervous system (Wierman & Crowley, 1986). The process is gradual and probably begins in middle childhood, long before any overt signs of sexual maturation are evident.

With regard to possible effects of training on the process of sexual maturation, several questions are relevant. Are there differential responses of the neuroendocrine system to training in the prepubertal and pubertal phases of growth and maturation? Can regular, intensive prepubertal training alter the hypothalamic gonadostat or neuroendocrine circuits, and in turn delay the adolescent spurt and sexual maturation? Are there possible cumulative effects of hormonal responses to regular training? On the other hand, can regular, intensive training during the pubertal phase of growth alter the progress of sexual maturation?

Girls

Discussions of training and sexual maturation most often focus on later mean ages at menarche in athletes. Distributions of ages at menarche in 176 white intercollegiate athletes and 228 other students at the same university are shown in Fig. 6.1. Later ages at menarche are more often associated with more advanced competitive levels, and intensive training is suggested as the factor which 'delays' menarche (Malina, 1983a).

A significantly later age at menarche implies later entry into reproductive life and thus a decrease in fertility potential. It is also suggested that an increase in age at menarche is associated with a relative decrease in fertility and increase in sterility (Behrman, 1967). In a prospective study of 2062 women, however, age at menarche was not related to total fertility and risk of stillbirth (Sandler, Wilcox & Horney, 1984). Risk of spontaneous abortion declined slightly with a later age at menarche across all pregnancies, but was not apparent when only first pregnancies were considered. In a retrospective study of the first four pregnancies in about 1000 women, spontaneous abortion rates also declined with an increasing age at menarche, but the trend was not affected by order of

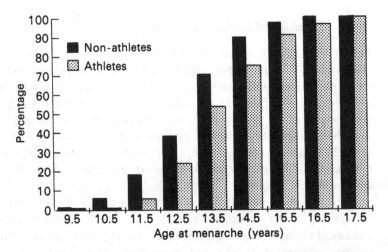

Fig. 6.1. Cumulative distribution of reported ages at menarche in athletes ($n = 176$) and non-athletes ($n = 228$). The athletes and non-athletes are all White and are enrolled in the same university (Malina, unpublished data).

pregnancy (Liestol, 1980). On the other hand, retrospective data for about 1000 women suggested a 'U-shaped' relationship between age at menarche and miscarriage rate in the first pregnancy: that is, miscarriage rates were highest in the earliest (<11 years) and latest (16+ years) maturers. Miscarriage rates declined with an increasing age at menarche from 11 through 14 years, and then increased among those attaining menarche at 15 and 16+ years (Martin, Brinton & Hoover, 1983). There was no trend for menarche and miscarriage rates in the second through fifth pregnancies.

Age at menarche is not related to age at menopause. This is shown for a survey in the Netherlands in Table 6.2 (Jaszmann, Van Lith & Zaat, 1969), while the correlation between age at menarche and age at meno-pause in women at the University of Minnesota is +0.05 (Treloar, 1974). There is a need for information on the age at menopause in former athletes and/or active women. Nevertheless, age of onset of the fertile span of life has a larger impact on total fertility than age of its termination.

Primary amenorrhoea is also apparently more common in athletes. Girls who have not menstruated spontaneously by 16 or 17 years of age are generally considered as having primary amenorrhoea (Behrman, 1967; Shangold, 1988). Using this criterion, 16 of 176 athletes (9%) and only 7 of 228 non-athletes (3%) attained menarche after their sixteenth birthday (Fig. 6.1) and would thus be classified as having primary

Table 6.2. *Relationship between age at menarche and age at natural menopause, Borough of Ede, The Netherlands**

n	Age at menarche (years)	Age at menopause (years)	
		Mean	SD
170	11	50.8	4.1
491	12	51.6	3.5
794	13	51.0	3.6
965	14	51.6	3.7
494	15	51.4	3.6
269	16	51.2	4.3
183	17–20	51.1	3.9

* Adapted from Jaszmann *et al.* (1969).

amenorrhoea. Although regular training is often implicated as the cause of primary amenorrhoea, the most common cause is constitutional delay (Shangold, 1988). In the survey of athletes, 3 of the total sample of 179 (1.7 %) indicated that they had never menstruated. One of the three had a congenital condition, while the clinical status of the other two was not known (Malina, unpublished data).

Data dealing with the inferred relationship between training and later menarche are associational and retrospective. A moderate correlation (+0.53) between years of training before menarche and age at menarche in runners and swimmers, for example, has been used to imply that training directly influences menarche: '... intense physical activity does in fact delay menarche' (Frisch *et al.*, 1981, p. 1562). Correlation, of course, does not imply a cause–effect sequence of events. The older a girl is at menarche, the more likely she would be to have begun her training prior to menarche, and conversely, the younger a girl is at menarche, the more likely it is that she would have had a shorter period of training before menarche (Stager, Robsertshaw & Miescher, 1984). More recently, computer simulation of quasi-experimental designs used in studies of the association between age at menarche and age at initiation of training before menarche indicates biased estimates of the statistical parameters (Stager, Wigglesworth & Hatler, 1990). This would suggest that the association between age at menarche and years of training before menarche is highly likely to be an analytical artifact. Further, later maturation may be a contributory factor in the selection process for sport rather than training *per se* causing the lateness (Malina, 1983*a*).

Menarcheal data are generally consistent with observations of breast and pubic hair development of young gymnasts and ballet dancers, though not of age-group swimmers (Malina, 1988). Given the popularity of running, it is somewhat surprising that the secondary sex characteristic development of young female runners has not been reported, but observations on skeletal maturation indicate a trend towards a delay (Malina, 1988; Seefeldt et al., 1988). Allowing for the largely cross-sectional nature of the available data, and selection practices in sports, especially as competitive levels increase, it is difficult to make inferences about regular training and the sexual maturation of young female athletes. Other factors which are known to influence menarche – for example, dietary restriction, sleep habits, family size and selection criteria for certain sports – should be controlled before making inferences about the effects of training.

It is also suggested that greater menarcheal delays occur in sport disciplines which emphasize low body weight. Low body weight, of course, may be the result of selection practices in specific sports and/or dieting. Warren & Brooks-Gunn (1989), for example, suggest that delayed menarche in young ballet dancers is related to their leanness which in turn may be associated with dieting during adolescence. Nevertheless, data for elite university level athletes indicate later menarche in athletes across several sports which differ considerably in emphasis on body weight: diving, track and field, swimming, volleyball, basketball, tennis and golf (Table 6.3).

A corollary of the suggestion that training delays menarche is that changes in body weight or composition associated with intensive training may function to delay menarche. This is related to the critical weight or critical fatness hypothesis that a certain level of weight (about 48 kg) or fatness (about 17 %) is necessary for menarche to occur (Frisch, 1976, 1988). Focus on relative fatness implies a role for the ratio of fat-free mass (FFM) to fat mass (FM). Accordingly, intensive training functions to reduce and maintain relative fatness below the hypothesized minimal level, or to increase FFM and thus raise the ratio of FFM to FM, thereby delaying menarche; in other words, maturation of young girls is delayed by keeping them lean through regular training, which, incidentally, has been suggested as a means of reducing the rate of teenage pregnancy (Frisch, 1981). Data on the effects of regular training on the FFM of girls are lacking, while data on relative fatness are more extensive (Bailey, Malina & Mirwald, 1986; Malina, 1983b). The hypothesis, however, may not apply to individuals experiencing emotional stress and to extremely muscular girls (Frisch, 1988). The critical weight or fatness hypothesis has

Table 6.3. *Mean ages at menarche in White athletes in several intercollegiate sports, 1985–1989, and White non-athletes at the same university**

		Age at menarche (years)	
Sport	n	Mean	SD
Basketball	19	13.85	1.03
Volleyball	23	13.77	1.65
Diving	16	13.87	1.59
Golf	20	13.76	0.99
Swimming	43	14.42	1.63
Tennis	19	13.95	1.59
Track and field	36	14.02	1.35
Sprinters	3	14.27	—
Middle-distance	4	15.14	—
Distance	15	14.01	1.31
Jumpers	8	13.88	1.10
Throwers	6	13.36	1.75
All athletes	176	14.02	1.45
Non-athletes, 1982	105	13.06	1.28
Non-athletes, 1987	123	12.96	1.41

* Malina (unpublished data). Data are also available for 44 Black athletes, distributed primarily in basketball (13) and the sprints (19). Mean age at menarche for the total group of Black athletes is 13.22 ± 1.60 years.

been discussed at length by many (Johnston *et al.*, 1975; Malina, 1978; Trussell, 1980; Scott & Johnston, 1982, 1985; Garn, LaVelle & Pilkington, 1983), and the evidence does not support the specificity of weight or fatness, or of threshold level, as the critical variable for menarche.

The suggested mechanism for the association between training and later menarche is hormonal. It is suggested that intensive training alters circulating levels of gonadotrophic and ovarian hormones, and in turn delays menarche. The hormonal data, however, are derived largely from studies of postmenarcheal women, both athletes and non-athletes. The majority of hormonal data do not deal with chronic changes associated with regular, intensive training; rather, they generally focus on acute responses to strenuous exercise.

The first sign of the onset of puberty is an increase in LH secretion during sleep late in childhood. The pulsatile release of LH is correlated with the number of sleep cycles in late prepubertal or early pubertal children and precedes the initial overt signs of puberty. The development of the nocturnal pulsatile pattern is apparently a critical aspect of sexual maturation. It is suggested that once this central nervous system mediated mechanism is initiated, it is resistant to change and is '. . . susceptible only to the most severe of influences, such as malnutrition or sleep deprivation' (Wall and Cumming, 1985, p. 76). As puberty progresses, the pulsatile release of LH gradually extends into the waking part of the day and the day–night differences in gonadotropin pulses become progressively less. With sexual maturity, LH secretion develops a cyclical pattern in females.

The pattern of LH production has not been considered in premenarcheal athletes engaged in regular training. In four postmenarcheal swimmers, Bonen et al. (1978) noted a short luteal phase. This same pattern, however, is also characteristic of teenage girls and may be a normal age-associated trend during maturation of the hypothalamic–pituitary–ovarian axis. Veldhuis et al. (1985) suggest that exercise-associated amenorrhoea in adult distance runners (which is discussed in more detail in a subsequent section) is due to a decreased pulsatile release of LH. It is not known, however, whether intensive training alters the pulsatile release of LH in premenarcheal girls who are in late prepubertal or early pubertal stages of development.

Caution is warranted in making inferences from postmenarcheal athletes to those in the process of sexual maturation. Growing and maturing individuals are not miniature adults and do not necessarily respond to environmental conditions, including acute or chronic exercise, in a similar manner to adults.

Hormonal data for active prepubertal or pubertal girls are limited. The results are not consistent across studies and the data base can be characterized as weak. The results are often based on single samples of hormones whose temporal sequence is markedly pulsatile. Low gonadotropin secretion in association with 'mild growth stunting', for example, has been reported in premenarcheal ballet dancers (Warren, 1980). The dancers were delayed in breast development, menarche and skeletal maturation, which would suggest a prolonged prepubertal state. They were not, however, delayed in pubic hair development.

Lower plasma levels of oestrone, testosterone and androstenedione have been observed in gymnasts compared to swimmers of the same age and prepubertal status. However, plasma gonadotropins and dehydroepiandrosterone-sulphate (DHEAS) levels did not differ between the two groups. On the other hand, plasma levels of these hormones did not differ between early pubertal (stage 2 of breast development: Tanner, 1962) gymnasts and swimmers, although the swimmers were, on average, older by about one-half of a year (Peltenburg *et al.*, 1984). Both the prepubertal and early pubertal gymnasts had been training regularly for a longer period than the swimmers, since 4.8 and 5.0 years of age, respectively, compared with 7.2 and 8.0 years, respectively. The similar levels of DHEAS in the prepubertal gymnasts and swimmers suggests a similar stage of adrenarche. Considering the difference in the duration of training in the two groups, this observation does not support the suggestion that training delays adrenarche and prolongs the prepubertal state (Brisson *et al.*, 1982). Moreover, more recent evidence does not support the view that secretion of adrenal androgens triggers sexual maturation (Wierman & Crowley, 1986).

The effects of regular training on basal levels of hormones in children and adolescents are not certain. Evidence for small samples of female swimmers 13–18 years of age indicates no differences in basal levels of oestradiol at the start and after 24 weeks of training in premenarcheal girls, but a lower basal level of oestradiol in postmenarcheal swimmers after 24 weeks. Both groups experienced a decline in basal levels during the first 12 weeks of training, followed by a rise at 24 weeks. The pre- and post-menarcheal swimmers also had similar basal levels of ACTH, cortisol, prolactin and testosterone during the 24-week training season (Carli *et al.*, 1983*a*). In the combined group, ACTH levels gradually increased, prolactin levels tended to increase, testosterone levels decreased, and cortisol levels showed a variable pattern during the season.

A role for beta-endorphins in amenorrhoea of runners and in turn delayed menarche in athletes has also been postulated (McArthur *et al.*, 1980). The effect of naloxone, an opiate receptor antagonist, under conditions of exercise in adults, for example, results in a marked increase in LH. However, responses of normal prepubertal and early pubertal (stages 1 and 2, respectively, of secondary sex characteristic development) girls and boys to naloxone under basal conditions is different from that of adults (Fraioli *et al.*, 1984). Naloxone did not have an effect on LH secretion in the children, even those in whom an LH response to exogenous LHRH was recorded. However, fundamental changes in the opiate neuroendocrine system may occur with the onset of puberty and may be related to its onset (Mauras, Veldhuis & Rogol, 1986).

Given the available information on the hormonal profiles of young athletes and the responses of young girls to training, it is difficult to implicate regular training as the critical factor in the later sexual maturation observed in many athletes. Constitutional delay is apparently the most common cause of later menarche observed in athletes (Shangold, 1988). Variation in family size may be a contributory factor (Table 6.4).

Table 6.4. *Estimated effects of family size on the age at menarche in samples of athletes and non-athletes*

Sample and reference	Estimated years per additional sibling
Athletes	
High school, US (Davison, 1981)	0.12
Olympic (Malina, 1983a)	0.22
University, US (Malina, unpublished)	
White	0.16
Black	0.17
Non-athletes	
Belgium, school girls (Wellens, 1984)	0.11
Great Britain	
School girls (Roberts et al., 1971)	0.18
School girls (Roberts et al., 1975)	0.15
University students (Roberts & Dann, 1967)	0.15
University students (Dann & Roberts, 1984)	0.12
Romania, school girls (Stukovsky et al., 1967)	0.17

For each additional sibling in a family, menarche is delayed by an estimated 0.11 to 0.22 years in several samples of athletes and non-athletes, or by about 1.5 to 2.5 months.

Boys

Possible relationships between systematic training and sexual maturation of boys has not been considered, although it has been suggested that '. . . boys may be better prepared physically for metabolic demands during the development of reproductive maturity . . .' (Warren, 1983: p. 370). In contrast, some literature suggests that the growth and maturation of males may be more susceptible to environmental alteration, while that of females may be better buffered against environmental stress (Bielicki & Charzewski, 1977; Stinson, 1985).

It is somewhat puzzling why one would expect training to delay the maturation of girls and not boys. The underlying neuroendocrine processes are generally similar, and other environmental stresses related to sport, such as stress and sleep, undoubtedly affect boys as well as girls. However, with the exception of wrestling, emphasis on extreme weight regulation is not characteristic of many sports for boys.

Data for young male athletes indicate a tendency towards early or advanced sexual and skeletal maturity status among participants in several team (baseball, American football) and individual (cycling, swimming, track and field) sports (Malina, 1988). Young ice hockey players (under 13 years of age), gymnasts, figure skaters (Malina, 1988) and distance runners (Seefeldt *et al.*, 1988), on the other hand, tend to be somewhat delayed in skeletal and/or sexual maturation.

Information on the hormonal responses of young males to acute or chronic exercise are not extensive. Emphasis is usually on testosterone and growth hormone, with a lack of data on LH. Among swimmers of 12–16 years of age (presumably pubertal), basal plasma levels of testosterone showed no change after 4 weeks of training, increased significantly after 12 weeks, remained rather stable between 12 and 24 weeks, and decreased below preseason values after 43 weeks of training and competition. In contrast, basal gonadotropin and growth hormone levels were not altered with swimming training. Most of the changes during the course of the season were variable and within the physiological range (Carli *et al.*, 1983*b*).

Hormonal responses to maximal exercise bouts are also variable among boys. For example, five young marathon runners, 13 years of age, experienced an increase in plasma testosterone relative to pre-race values, while 17- and 40-year-old runners experienced a significant depression in plasma testosterone after the race (Schmitt, Schnabel & Kindermann, 1983). In contrast, postmaximal exercise (cycle ergometry) concentrations of serum testosterone did not differ in boys grouped by stage of pubertal development (Fahey *et al.*, 1979). On the other hand, relative increases in plasma testosterone after incremental treadmill exercise to $\dot{V}_{O_2}$max were greater in a longitudinal study of prepubertal and late-pubertal boys, but responses of mid- and post-pubertal boys were more variable (Wall & Cumming, 1985).

Overview

Studies of the hormonal responses of young athletes of both sexes are variable and not conclusive. Acute changes in response to exercise and chronic changes associated with regular, intensive training must be distinguished. Many studies are based on single serum samples, but virtually all hormones are episodically secreted. Studies in which 24-hour levels of hormones are monitored or in which actual pulses of hormones are sampled every 20 minutes or so are needed. Further, the simple presence of a hormone does not necessarily imply that it is physiologically active. There may be variation in the responsiveness of hormone receptors at the tissue level. It is possible that training may have an influence on hormonal receptors at different stages of pubertal development, but this issue has not yet been addressed.

Since sleep apparently has an important role in the maturation process in both sexes, the effects of training and competition on sleep in young athletes merit closer scrutiny. The effects of daytime exercise on sleep – longer sleep, more slow-wave sleep, and more growth hormone production during sleep – are presumably beneficial (Oswald, 1980; Weitzman, 1980). On the other hand, stress is associated with both increased and decreased LH concentrations in the waking state (Shangold, 1984). The effects of pre- and post-competition anxiety on sleep and hormonal production might thus be a seemingly contradictory factor in late prepubertal or early pubertal athletes.

Menstrual dysfunction and training

The prevalence of secondary amenorrhoea (cessation of menses) and oligomenorrhoea (infrequent menses) is more common among athletes, including ballet dancers, than the general population. Estimated prevalence varies from about 10 % to 50 % among various samples of athletes compared to about 2 % to 5 % in the general population (Baker, 1981; Drinkwater, 1984; Loucks & Horvath, 1985; Shangold, 1988). The data are largely retrospective and based on questionnaires. The different estimates are related in part to variable criteria for amenorrhoea and oligomenorrhoea. In addition, many athletes are of college age, and college women in general have a higher prevalence of secondary amenorrhoea and oligomenorrhoea than the general population (Bachmann & Kemmann, 1982). This is probably related in part to emotional stress associated with campus life – for example, examinations, social interactions, dietary changes, and so on – and in part to physiological immaturity, since the hypothalamic–pituitary–ovarian axis in some college women is probably not yet fully mature. Regularly menstruating non-athletic college women, for example, have lower levels of luteal progesterone than observed in older women (Ellison, Lager & Calfee, 1987), which would suggest a degree of immaturity or a prolonged process of ovarian maturation.

Menstrual dysfunction occurs more commonly in runners and ballet dancers than in athletes in other sports, who nevertheless experience a higher prevalence than the general population. However, it is apparently not related to running pace, average number of miles run per week and years of training among runners, although results of several surveys are variable (Loucks & Horvath, 1985; Shangold, 1988).

Table 6.5. *Factors associated with menstrual dysfunction in athletes*

1 Late (delayed) menarche
2 Prior menstrual dysfunction
3 Nulliparity and young age
4 Physical and emotional stress of training/competition
5 Weight loss
6 Changes in body composition, ratio of fat to lean tissue
7 Dietary changes, nutritional inadequacy
8 Acute effects of exertion
9 Chronic effects of repeated exertion
10 Hormonal alterations

Compiled from Baker (1981), Cumming & Rebar (1983), and Loucks & Horvath (1985).

Table 6.6. *Menstrual patterns of participants in the 1979 New York City Marathon prior to and during training*

		Menstrual pattern during training				
		Regular		Irregular		Amenorrhoeic
Menstrual pattern prior to training*	n	n	%	n	%	n	%
Regular	270	251	93	15	6	4	1
Irregular	54	14	26	37	69	3	6
Amenorrhoeic	6	1	17	0	0	5	83

Adapted from Shangold & Levine (1982).
* Regular was defined as consistent intervals of 23 to 36 days; irregular was defined as inconsistent intervals of <23 or >37 days and <6 months; amenorrhoeic was defined as no more than one bleeding episode in the previous 10 months.

It is important to note that not all athletes and not all women who train experience oligomenorrhoea or amenorrhoea: that is, not all women are equally susceptible. Hence, factors which are related to or perhaps which dispose certain women to menstrual dysfunction need to be considered (Table. 6.5).

Menstrual history

Some evidence suggests that menstrual function during training is dependent in part on menstrual function prior to training. This is shown in Table 6.6 for a sample of participants in a marathon who completed detailed gynaecological questionnaires (Shangold & Levine, 1982). In general, runners who had a regular menstrual pattern prior to training tended to have a regular menstrual pattern during training, while runners who had an irregular menstrual pattern prior to training tended to have an irregular pattern during training. Five of the six amenorrhoeic runners prior to training remained amenorrhoeic. Only a small percentage of runners with regular or irregular patterns prior to training became amenorrhoeic during training, but relatively more runners with irregular menstrual patterns became amenorrhoeic. In a comprehensive survey of a small sample of runners, Schwartz et al. (1981) also noted that those with amenorrhoea had a higher prevalence of prior menstrual irregularity than other runners and non-runners.

Age and parity

Several surveys suggest that age and prior pregnancy may influence the development of menstrual dysfunction during training. Among marathon participants, nulliparous runners reported a higher prevalence of oligomenorrhoea and amenorrhoea during training (19/37, 51 %) than parous runners (11/52, 21 %) (Dale, Gerlach and Wilhite, 1979). However, age was not controlled in the comparison. In another survey of marathon runners, the majority who reported secondary amenorrhoea during training ($n = 53$) were under 25 years of age (Speroff & Redwine, 1980). In a small sample of distance runners ($n = 23$), a nulliparous group ($n = 15$) was younger and included a greater percentage of amenorrhoeic runners (Baker et al., 1981). The amenorrhoeic runners had a later age at menarche, which led the authors to postulate that secondary amenorrhoea in younger runners may be related to later age at menarche. Association, of course, does not imply a cause–effect sequence of events.

Weight loss and body composition

Weight loss and changes in body composition during training are probably related. Much of the discussion focuses on relative fatness in the context of Frisch's (1976) hypothesis that 22 % body fat is necessary to maintain regular ovulatory menstrual cycles. The data, however, do not support the hypothesis. Part of the problem relates to the method of estimating relative fatness in the original hypothesis: total body water is predicted from stature and weight, and relative fatness is subsequently derived. These procedures have severe statistical limitations (Reeves, 1979; Trussell, 1980) and tend to overestimate the fatness of athletes (Malina, 1983a). Further, in many subsequent studies, body density is often predicted from skinfolds, a procedure with well-documented limitations.

Nevertheless, changes in body composition are not the primary factor in menstrual dysfunction. For example, among 14 eumenorrhoeic and 14 amenorrhoeic athletes matched for sport (11 runners and 3 crew members of each group), age, stature, weight and training regimes, Drinkwater *et al.* (1984) reported no significant differences in densitometrically estimated relative fatness (16.9 ± 0.8 % and 15.8 ± 1.4 %, respectively) and fat-free mass (48.0 ± 1.6 kg and 45.6 ± 1.6 kg, respectively). McArthur *et al.* (1980) reported similar estimates of relative weight and relative fatness in three amenorrhoeic runners and six control subjects. Further, Warren (1980) noted that a reduction in training among amenorrhoeic ballet dancers without significant changes in relative fatness or body weight was apparently sufficient for the resumption of menstrual cycles. These results would thus suggest that relative fatness and relative weight *per se* are not the primary factors in menstrual dysfunction observed in runners and other athletes.

Diet and eating disorders

Dietary changes associated with training are perhaps related to menstrual function. It is suggested that training in association with an 'energy drain' may contribute to menstrual irregularity and amenorrhoea (Warren, 1980). However, detailed dietary analyses in conjunction with estimates of energy expenditure are not extensive for eumenorrhoeic and amenorrhoeic athletes. Using prospective 7-day dietary records, Schwartz *et al.* (1981) reported that both amenorrhoeic runners ($n = 12$) and eumenorrhoeic distance runners ($n = 34$) consumed more calories than eumenorrhoeic middle-distance runners ($n = 24$) and control subjects ($n = 20$). It is of interest that the amenorrhoeic and eumenorrhoeic long-distance runners consumed, on average, only about 1800 kcal/day while running, respectively, about 40 miles/week and 48 miles/week, which would lead one to question whether their energy intake was adequate to meet their energy needs. Fat-free mass is, of course, a source of energy when caloric needs are not met in the diet. The amenorrhoeic runners also had a lower percentage of daily intake of protein compared with other runners and controls. Kaiserauer *et al.* (1989) noted lower consumption of fat, red meat and total calories in amenorrhoeic ($n = 8$) compared to regularly menstruating distance runners ($n = 9$). The authors suggest that nutritional inadequacy may distinguish amenorrhoeic runners from regularly menstruating runners, and that training alone may not be the major factor associated with amenorrhoea in distance runners.

Eating disorders may be a confounding factor in menstrual dysfunction among athletes. Patients with anorexia nervosa have a high prevalence of amenorrhoea (Blumenthal, O'Toole & Chang, 1984), while a significant percentage of athletes show one or more potentially pathogenic weight control behaviours, such as binge eating, self-induced vomiting, and use of diet pills, laxatives and/or diuretics (Rosen et al., 1986). Athletes in sports which emphasize leanness are apparently those who more often show considerable preoccupation with body weight and a tendency towards eating disorders (Borgen & Corbin, 1987). Further, it has been suggested that obligatory or compulsive runners share personality traits with anorectic patients, especially in their preoccupation with food and obsession with leanness (Yates, Leehey & Shisslak, 1983). However, using the Minnesota Multiphasic Personality Inventory, Blumenthal et al. (1984) did not observe either a common set of personality traits or the same degree of psychopathology in obligatory runners and patients with anorexia nervosa.

Data on eating disorders among athletes with menstrual dysfunction are not extensive. Among classical ballet dancers with menstrual irregularity, for example, Calabrese et al. (1983) reported a high prevalence of binge eating (70 %). Mean estimated daily energy intake, however, was well below recommended dietary allowances at 1360 kcal/day. Hamilton et al. (1988) noted that highly selected ballet dancers (those in companies who chose dancers exclusively from company schools) have a lower prevalence of eating problems and anorectic behaviours than less selected dancers (those selected from general auditions). The authors suggest (Hamilton et al., 1988: p. 564) that

> ... dancers who have undergone a stringent process of early selection may be less susceptible to the development of eating problems because they are more naturally suited to the thin ideal required by this profession.

Stress

Psychological and emotional stress is also related to menstrual dysfunction. Responses to stress, however, are variable. For example, among 91 postmenarcheal girls at a boarding school, 42 % showed alterations in the menstrual cycle during examinations (Dalton, 1968). The cycle was longer in 19 and shorter in 11 girls, while 8 had missed their normal period (temporary amenorrhoea) during examinations. Other stresses such as death of a family member or friend and other distressing personal experiences are also associated with temporary menstrual dysfunction (Prior, 1982). The impact of the psychological and emotional stress of training and competition, probably in concert with the physical demands, on menstrual dysfunction of athletes has not been carefully considered. It is of interest that amenorrhoeic runners subjectively associated more stress with running than did eumenorrhoeic long- and middle-distance runners, even though the groups did not differ in psychological profiles designed to measure depression, hypochondriasis, obsessive/compulsive tendencies and recent stressful events (Schwartz *et al.*, 1981). Significant percentages of obligatory runners ($n = 43$) indicated feelings of guilt if they did not run (86%) and/or feeelings of tension, irritability or depression if they missed a training session (72%) (Blumenthal *et al.*, 1984).

Stress, of course, is associated with temporary hormonal and metabolic changes. Repeated or prolonged stress, however, may have long-term effects, including perhaps a direct effect on the neuroendocrine system and, in turn, menstrual dysfunction.

Hormonal changes

Specific mechanisms underlying menstrual dysfunction are not known with certainty, but alterations in the neuroendocrine system are involved. Short-term aerobic exercise induces increases in plasma and serum levels of gonadotrophic and ovarian hormones and in endogenous opiates, though some of the data for gonadotropin levels are inconsistent (Shangold, 1984, 1988). There is a wide range of variation in hormonal responses among individuals. The increases are largely transient, probably reflecting responses to the demands of acute exercise. Hormonal responses to repeated bouts of exercise such as those in regular training are more relevant to menstrual dysfunction.

Hormonal studies have several methodological limitations: for example, repeated sampling during the course of a day is necessary to establish baseline values and changes during training, daily sampling may be required during a training programme, and so on. More importantly, virtually all hormones are secreted episodically and have distinct diurnal patterns. Many studies are based on single plasma samples and thus have limitations, in that the secretory pulse of the hormone in question may have been missed. Further, plasma levels of hormones are not as simple as their values would suggest. They '. . . represent a balance among

Table 6.7. *LH pulse frequency in adult controls and adult runners, with information on age at menarche and training history for the runners*

LH pulse frequency (pulses/24 h)	Adult controls	Adult runners		
		Subject	Age at menarche (years)	Training history (years)
15	X	V	13	4
14				
13		VI	17	10
12		IV	14	5
11	XXXX			
10	XX			
9	XX			
8	X			
7				
6		IX	16.5	15
5		VII	13	1.5
4		VIII	12	3
3		II	14	2
2		I	12	6
1		III	14	5
Group mean ± SEM	10.5 ± 0.6	6.8 ± 1.7		

Adapted from Veldhuis *et al.* (1985).

production, metabolism, utilization, clearance, and plasma volume – all of which may change simultaneously with exercise' (Shangold, 1988: p. 135). Other factors can also influence hormonal levels, including state of sleep or wakefulness, diet, state of feeding or fasting, emotional states, body weight and composition (Shangold, 1984, 1988).

Allowing for limitations of hormonal studies, some data emphasize a significant role for impaired LH production in menstrual dysfunction observed in athletes (see Loucks & Horvath, 1985). As noted earlier, actively training swimmers have shorter luteal phases (Bonen *et al.*, 1978). Following the menstrual function of a healthy, ovulatory 30-year-old runner during 18 cycles, Shangold *et al.* (1979) noted shorter luteal phases in cycles which had greater running mileage. Changes in the luteal phase in association with intensive training have subsequently been reported by others (see Shangold 1984, 1988).

Veldhuis *et al.* (1985) suggests that exercise-associated amenorrhoea in adult distance runners is due to a decreased pulsatile release of LH. In this study, six out of nine amenorrhoeic runners showed a decrease in pulsatile LH secretion (Table 6.7). Note, however, that not all women experienced altered LH secretion with training. Two of the nine runners began regular exercise before menarche and had later menarche, 16.5 and 17 years of age, respectively, while the seven other runners attained menarche between 12 and 14 years. It is of interest that one of the two runners with early onset of training and later menarche had a LH pulse frequency within the range of control subjects, while the other had a LH pulse frequency just below the control range. The five other runners with reduced LH pulses had ages at menarche between 12 and 14 years.

Bullen *et al.* (1985) reported results of a prospective study of 28 previously untrained women with regular ovulatory cycles and luteal adequacy who were exposed to a programme of running and moderate intensity sport activities over the span of two cycles. The women were assigned to weight-maintenance and weight-loss groups. Only four women did not show menstrual abnormalities, and the prevalence of abnormalities was greater in the weight-loss group. Hormonal tests (abnormal luteal function, loss of LH surge) identified a greater percentage of menstrual abnormalities than clinical criteria (abnormal bleeding, delayed menses). The weight-loss group experienced a higher frequency of abnormal luteal function during the first part of the training programme, and a loss of the LH surge as training progressed. In contrast, abnormal luteal function and loss of LH surge did not vary during the training programme in the weight-maintenance group.

In contrast, a less strenuous exercise programme with seven college women who were not involved in systematic training did not disturb menstrual cycles (Bullen *et al.*, 1984). Hormonal tests, however, indicated disturbed ovarian function in four subjects. Results of the preceding two studies indicate the importance of hormonal tests for documenting menstrual changes in response to training and other stimuli. Commonly used clinical criteria and self-reporting may not be sufficiently specific.

The results also suggest an apparent training threshold effect. More intensive training apparently has a greater effect on the hypothalamic–pituitary–ovarian axis than moderately intensive programmes. Preliminary results from a prospective study of initially untrained women training to run a marathon are consistent with a threshold effect (Rogol *et al.*, 1989). Shortened luteal phases occurred only in those women who trained above the lactate threshold.

A role for beta-endorphins in the amenorrhoea of runners has been postulated (McArthur *et al.*, 1980). Since administration of naloxone, an opiate receptor antagonist, resulted in a marked increase in LH in one of three amenorrhoeic runners, it was suggested that increased levels of beta-endorphins during exercise may lead to reduced secretion of GnRH and, in turn, suppressed levels of LH. However, similar responses to naloxone were not observed in four amenorrhoeic runners (Cumming & Rebar, 1983). Among 14 hypothalamic amenorrhoea patients, on the other hand, naloxone resulted in increased LH pulse frequencies (Khoury *et al.*, 1987). There is thus the possibility that the amenorrhoea observed in runners and other athletes may be a distinct clinical entity in contrast to hypothalamic amenorrhoea.

Overview

The preceding discussion emphasizes the complexity of factors associated with menstrual dysfunction in trained women. It should be noted that some women who do not train also experience these problems. Training is thus only one factor, and in many studies it is not adequately quantified. Several factors related to menstrual problems are components of the social environment. Hence, menstrual dysfunction cannot be approached in a solely biological manner. A biosocial perspective is essential, recognizing the interaction of biological and social factors.

Nevertheless, alterations in the hypothalamic–pituitary–ovarian axis occur in menstrual dysfunction. The specific mechanism(s) is (are) not known with certainty. Data on changes in LH pulse frequency would seem to suggest a hypothalamic origin. It has also been suggested that peripheral changes are of primary importance – for example, changes in circulating ovarian hormones and/or alterations in clearance rates of gonadotropins and ovarian steroids – and these secondarily affect the hypothalamic–pituitary–ovarian axis. There is also the possibility that higher brain centres may inactivate a functionally patent neuroendocrine axis.

Implications of menstrual dysfunction

What is the long-term effect of sustained oligomenorrhoea and/or amenorrhoea on the reproductive system? Does the hypothalamic–pituitary–ovarian axis need to be stimulated on a regular basis to maintain functional integrity? Does menstrual dysfunction have a permanent effect on the reproductive system? What is the relationship of menstrual dysfunction to fertility?

It is generally believed that oligomenorrhoea and amenorrhoea associated with regular training are reversible, although '. . . this reversibility has not been completely proven' (Baker, 1981: p. 694). When training is interrupted, many amenorrhoeic athletes resume menses, quite often with no major change in body weight. In the study of Bullen *et al.*, (1985) cited above, all of the women resumed normal menstrual cycles within 6 months after the training programme ceased. Further, observations on elite athletes in a variety of sports indicate no significant effect of early training on subsequent reproductive function (see below). Many of these studies, however, do not consider menstrual dysfunction in the athletes.

Luteal changes associated with amenorrhoea can lead to luteal phase deficiency which is associated with infertility (Shangold, 1988). Luteal phase deficiency is associated with deficient secretion of progesterone. It is interesting that lower luteal salivary progesterone levels are associated with even moderate recreational running (Ellison & Lager, 1986). Luteal phase defects, however, are apparently reversible. For example, one of two distance runners who had luteal phase deficiency had a normal pregnancy when she stopped running (Prior *et al.*, 1982).

Although not related to training *per se*, a recent survey of infertile women with ovulatory failure considered their exercise histories (Greene *et al.*, 1986). Vigorous exercise for an average of less than 1 hour per day did not appear to be related to primary or secondary infertility. However, a history of vigorous exercise for 1 hour or more per day occurred more commonly among nulligravid cases. Nevertheless, other factors must be considered. Age at marriage and sexual practices may be as important in respect to total fertility as are biological components, which would emphasize the need for a biocultural approach in considering the implications of menstrual dysfunction in some athletes and active women.

The long-term consequences for fertility of menstrual dysfunction associated with training thus merit closer examination. According to Shangold (1984: p. 71):

> It is more likely that runners with reproductive dysfunction are less able to become pregnant, while those with normal reproductive function are better able to become pregnant. Although infertility is reported to be no higher for runners than for the general population, this may be true because many women runners who are actually infertile have not yet tested their fertility, and therefore are unaware that they are infertile.

Training and pregnancy

Maternal exercise during pregnancy is an issue that is presently receiving more attention. Pregnancy places a number of physiological demands on the mother, including, for example, greater energy and nutrient requirements, greater oxygen consumption, and increases in cardiorespiratory function. The developing individual, of course, needs energy, nutrients and oxygen for his/her metabolism and growth. Some concern has thus been expressed about the possibility of compromising the oxygen, energy and nutrient needs of the developing foetus and perhaps affecting foetal outcome through exercise stress on the mother.

A related issue is the course and outcome of pregnancy in individuals with a history of regular, intensive training, such as former athletes. Athletes who regularly trained during childhood and adolescence probably experienced some degree of developmental acclimatization to the rigours of training. This population is probably different from women who began recreational exercise such as running or aerobics as adults without a prior history of regular training. A good deal of the literature on training and pregnancy is based on the former.

Training during pregnancy

In one of the more comprehensive studies, Clapp & Dickstein (1984) considered the effects of endurance exercise (running and aerobic dance) on the outcome of pregnancy in a prospective study of 336 women. All information on exercise performance was obtained by interview. The women were divided into three groups:

Group 1: no regular exercise prior to and during pregnancy;
Group II: regular endurance exercise at or above minimum conditioning levels prior to pregnancy, reduced level of exercise early in pregnancy and stopped exercising completely prior to the twenty-eighth week; and
Group III: regular endurance exercise at or above minimum conditioning levels prior to pregnancy and maintenance of exercise habits at or near prepregnancy levels throughout pregnancy into the third trimester.

Comparisons of several indicators of pregnancy outcome in the total sample and subsamples matched for age, parity, prepregnancy weight and social class are shown in Table 6.8. Groups I and II did not differ significantly in any of the parameters, which suggests that endurance exercise prior to pregnancy does not influence pregnancy weight gain, gestational length and birthweight. However, continued endurance exercise during pregnancy is associated with reduced weight gain, shorter gestational length and lower birthweight. Complications of pregnancy

Table 6.8. *Prepregnancy weight and pregnancy outcome (mean ± SEM) relative to endurance exercise status (see text for exercise characteristics of each group)*

Total sample	Group I (n = 152)	Group II (n = 47)	Group III (n = 29)
Prepregnancy weight (kg)	61.9 ± 1.1	59.7 ± 1.2	59.0 ± 2.0
Pregnancy weight gain (kg)	14.6 ± 0.4	16.8 ± 0.8	12.2 ± 0.6
Gestational length (days)	280.7 ± 1.4	281.8 ± 2.2	273.7 ± 2.2
Birthweight (g)	3518 ± 43	3577 ± 92	3009 ± 100

Samples matched for age, parity, prepregnant weight and socioeconomic class

	Group I (n = 29)	Group II (n = 29)	Group III (n = 29)
Pregnancy weight gain (kg)	16.0 ± 0.7	17.9 ± 1.1	12.2 ± 0.6
Gestational length (days)	280.2 ± 1.5	286.5 ± 2.3	273.7 ± 2.2
Birthweight (g)	3632 ± 77	3686 ± 81	3009 ± 100

Adapted from Clapp & Dickstein (1984).

Table 6.9. *Prepregnancy weight and pregnancy outcome (mean ± SEM except as indicated) in women who regularly performed endurance exercise prior to pregnancy and maintained their exercise at or near preconceptual levels into the third trimester, grouped by level of endurance exercise (Group III in Table 6.8)*

	Level of endurance exercise		
Outcome variable	High (n = 6)	Medium (n = 13)	Minimum (n = 10)
Prepregnancy weight (kg)	57.8 ± 3.3	55.5 ± 1.4	61.5 ± 2.5
Pregnancy weight gain (kg)	10.8 ± 0.8	11.8 ± 1.4	12.7 ± 2.5
Gestational length (days)	265.4 ± 3.5	274.3 ± 5.9	276.2 ± 3.0
Birthweight (g)	2633 ± 252	3034 ± 173	3155 ± 99
Birthweight (range)	1764–3550	1820–4040	2700–3750
Birthweight <10th percentile (n)	4	5	2

Adapted from Clapp & Dickstein (1984).

and of the neonatal period were generally low and evenly distributed among the three groups.

The women in Group III were further considered relative to the intensity of exercise during pregnancy (Table 6.9). The results suggest a dose–response effect. Although numbers are small, women who engaged in high-level endurance exercise well into the third trimester experienced, on average, less weight gain and shorter gestational lengths and their infants had lower birthweights for gestational age compared to those who exercised at medium and minimum levels. There was, however, no increase in immediate morbidity.

Results of several studies with smaller samples are not entirely consistent with the preceding. There are, however, differences in design and intensity of the exercise stress. Pomerance, Gluck & Lynch (1974) and Erkkola & Makela (1976), for example, reported no relationship between maternal level of physical fitness and gestational length, birthweight and Apgar score. Erkkola (1976) also reported no disturbances of pregnancy associated with near-maximal exercise tests during various stages of pregnancy. Dale, Mullinax & Bryan (1982) observed no significant differences in weight gain during pregnancy, length of labour and obstetrical complications between 12 runners who ran regularly into the third trimester and 11 sedentary controls. The authors suggest, however, an increased Caesarean section rate in the runners. Birthweights of infants born to the runners and controls also did not differ. Collings, Curet & Mullin (1983) reported no differences in duration of labour and in the birthweights and Apgar scores of infants born to mothers who exercised on a cycle ergometer at 65 % to 70 % of predicted maximal oxygen consumption during the second and third trimester ($n = 12$) and controls ($n = 8$). However, though the results were not significant, the infants born to the mothers who exercised had, on average, greater weights and lengths and placental weights. Finally, Berkowitz et al. (1983) observed no relationship between physical activity during pregnancy (employment, housework, child care and leisure-time physical activity) and risk of pre-term delivery. Rather, moderate leisure-time physical activity was associated with reduced risk of pre-term delivery.

Thus, mild-to-moderate physical activity during pregnancy, particularly during the first 6 to 7 months, apparently has no effect on foetal development. The evidence for prolonged strenuous exercise suggests lower birthweights, which is consistent with experimental data from several species of animals (see Gorski, 1985; Lotgering, Gilbert & Longo, 1985).

Pregnancy in athletes

Observations on elite athletes in a variety of sports indicate no significant effect of early training on subsequent reproductive function (Niemineva,

Table 6.10. *Means and ranges for ages at menarche and at first parturition in East German athletes*

		Age at menarche			Age at first parturition	
Sport	*n*	Mean (years)	Range (years)	*n*	Mean (years)	Range (years)
Handball	98	13.0	10.1–15.0	25	23.0	19.1–28.1
Canoeing	32	13.0	10.1–15.1	8	22.1	19.1–24.1
Volleyball	63	13.1	11.1–15.0	20	22.0	18.0–24.1
Swimming	52	13.1	11.1–15.0	18	23.1	18.1–25.0
Athletics	102	13.1	11.0–16.1	107	23.1	18.0–29.1
Diving	26	14.0	12.1–17.1	18	23.1	19.1–28.0
Figure skating	30	15.0	12.1–19.0	6	23.1	20.0–26.0
Gymnastics	25	15.0	13.0–19.0	17	22.1	19.1–24.1

Adapted from Marker (1983).

1953; Erdelyi, 1962; Astrand *et al.*, 1963; Zaharieva, 1972; Eriksson *et al.*, 1978; Marker, 1983). Disorders of pregnancy, duration and course of labour, and obstetric complications are apparently no more common in athletes than in non-athletes, and in some instances less in athletes. Ages at menarche and at first parturition in East German athletes are summarized in Table 6.10. Over 80 % of the athletes had been actively training for 10 years or more. Mean ages at first parturition do not differ among athletes in several sports, although mean ages at menarche vary. This would suggest that training during childhood and/or adolescence does not influence reproductive performance. If early training has an effect on menarche and subsequent menstrual function, it is apparently temporary.

The distribution of birthweights of infants born to formerly and presently active athletes is generally similar to appropriate reference data (Niemineva, 1953; Marker, 1983). For example, among 242 infants born to East German athletes, only 6.6 % had birthweights less than 2500 g (Marker, 1983). Corresponding figures for children born to North American white women of 20–24 and 25–29 years of age respectively in 1975 are 6.0 % and 5.4 % (National Center for Health Statistics, 1977).

Sex hormones in trained males

In contrast to the rather extensive literature on the effects of training on the reproductive function of women, there is a paucity of corresponding

172   R. M. Malina

information for men. Circulating levels of testosterone generally increase during submaximal and maximal exercise, but decline with prolonged exercise of more than 3 hours (Sutton et al., 1973; Dessypris, Kuoppa-salmi & Aldercreutz, 1976; Galbo et al., 1977; Gawel et al., 1979; Guglielmini, Paolini & Conconi, 1984; Wall & Cumming, 1985). Individual variation in hormonal responses is considerable. Elevated testosterone levels are not generally accompanied by changes in levels of LH (Sutton et al., 1973; Dessypris et al., 1976; Galbo et al., 1977; Wall & Cumming, 1985). Further, the frequency of LH pulses is apparently not influenced by acute exercise in male distance runners (MacConnie et al., 1986; McColl et al., 1989). The physiological significance of elevated testosterone levels with exercise is not certain. However, some evidence suggests that the increase may be a function of a reduction in the rate of clearance and not due to an increase in testosterone production (Cadoux-Hudson, Few & Imms, 1985; Wall & Cumming, 1985). The effects of training on the binding, production and clearance of testosterone need further investigation.

Chronic effects of regular endurance running on testosterone levels are different from responses to acute exercise, and parallel, in general, those observed in women. Fourteen of 20 marathon runners presented significantly depressed levels of total testosterone, but levels of free testosterone were within the physiological range (Ayers et al., 1985). Among 31 runners training more than 64 km/week, only four had serum levels of total testosterone that were above the mean levels for controls and one had a value below the normal physiological range (Wheeler et al., 1984). Levels of total testosterone, free testosterone, non-specifically bound testosterone and prolactin of runners were, on average, significantly below those of control subjects, but values were within the physiological range. In a related study from this laboratory (McColl et al., 1989), six runners, who averaged 80 km per week, had a significantly lower level of serum testosterone, lower LH pulse amplitude and reduced area under the LH curve compared to six age-matched, sedentary controls at rest. On the other hand, the runners and controls did not differ in LH pulse frequency. Thus, in contrast to studies of female runners, who show a reduction in both LH pulse amplitude and frequency, these data for male runners suggest that endurance training in males results in a reduction of only LH pulse amplitude and not of LH pulse frequency.

Among six runners with greater training distance (125–200 km/week), plasma concentrations of testosterone, LH and FSH were, on average, similar to those in 13 age-matched controls (MacConnie et al., 1986). However, the frequency and amplitude of LH pulses at base line were

Table 6.11. *Plasma concentrations (mean ± SEM) of gonadotropins and testosterone at base line in male marathon runners (n = 6) and controls (n = 13)*

	Runners	Controls
LH (mIU/ml)	2.4 ± 0.05	2.5 ± 0.06
LH pulse frequency (n/8 h)	2.2 ± 0.48	3.6 ± 0.24
LH pulse amplitude (mIU/ml)	0.9 ± 0.24	1.6 ± 0.15
FSH (mIU/ml)	1.7 ± 0.08	1.6 ± 0.07
Testosterone (ng/ml)	6.2 ± 0.46	6.0 ± 0.36

Adapted from MacConnie *et al.* (1986).

Table 6.12. *Responses (mean ± SEM) of plasma LH to exogenous doses of GnRH in male marathon runners and controls*

Response	Dose of GnRH (µg/kg)		
	0.25	0.50	2.50
Peak response (mIU/ml)			
Runners	6.0 ± 0.86	8.9 ± 1.98	12.0 ± 3.31
Controls	8.0 ± 1.66	12.2 ± 3.28	17.7 ± 5.60
% increase			
Runners	154 ± 30.9	270 ± 65.8	394 ± 112.6
Controls	349 ± 57.3	555 ± 119.6	858 ± 228.2

Adapted from MacConnie *et al.* (1986).

significantly lower in the runners than in the controls (Table 6.11), which is in contrast to the observations of McColl *et al.* (1989) for males training about 80 km/week. Moreover, when exogenous GnRH was administered, the runners had a reduced LH response at all doses (Table 6.12).

Results of the preceding studies thus suggest possible alterations in the function of the hypothalamic–pituitary–gonadal axis in endurance-trained males which are similar to those observed in trained women with menstrual dysfunction. However, the observations on males may suggest a threshold effect of endurance training on LH pulse frequency. Runners with a training volume of about 80 km/week show normal pulsatile release of LH at rest, but those with a training volume of 125–200 km/week show decreased pulsatile release of LH at rest.

It is likely that several factors, in addition to regular endurance training, are involved in the altered neuroendocrine function observed in highly trained males. Eating disorders occur among males, and it has been suggested that some marathon runners show anorectic behaviours, such as an obsession with leanness, marked weight loss and elevated stress (Ayers et al., 1985). Some collegiate wrestlers experience significant reductions in serum testosterone and prolactin concentrations, though not in LH and FSH, during the season (Strauss, Lanese & Malarkey, 1985). The lowest testosterone levels were associated with low relative fatness, leading the authors to suggest that the depressed testosterone levels may reflect the consequences of severe dietary restriction.

Implications of altered reproductive function for fertility in endurance trained males is not known, although depressed spermatogenesis may be possible. Two of 20 marathon runners, for example, presented severe oligospermia and both had the lowest levels of total and free testosterone (Ayers et al., 1985).

Physical activity and longevity

The role of habitual physical activity in longevity is complex. Habitual physical activity can influence longevity through an effect on various disease processes, especially degenerative diseases of the cardiovascular system. Physical inactivity is a risk factor in cardiovascular disease morbidity and mortality. In a comprehensive review of 43 studies, Powell et al. (1987) concluded that there is a direct relationship between a sedentary life-style and coronary heart disease and that the relationship suggests a dose–response gradient. Although the evidence is less substantial and there are many confounding factors, physical inactivity may be a risk factor for morbidity and mortality from several cancers and adult-onset diabetes in both sexes (Blair, 1988).

Studies of the longevity of male athletes yield mixed results. Among 17 studies comparing athletes to the general population, 16 suggest lower mortality and increased longevity among athletes and 2 emphasize an important role for calibre of the athlete and level of competition, both of which are related (Stephens et al., 1984). As noted earlier, athletes are a highly select group, not only for skill, but for other characteristics, including physique, socioeconomic status and level of education. These factors may confound population-based comparisons and stress the need for adequate control groups.

A variety of studies compares longevity of male university athletes to control groups who are usually university classmates. The results, however, are inconsistent among studies (Stephens et al., 1984). Several emphasize the need to consider sport-specific variation in longevity and physique. Among Harvard athletes, the longevity of letter winners in major varsity sports of the period (baseball, football, crew, track, ice hockey and tennis) was less than that of minor athletes (non-letter winners in the major sports and participants in minor sports of the period, such as basketball, cricket, fencing, golf, lacrosse, polo, swimming and wrestling) (Polednak & Damon, 1970). The longevity of minor athletes was greater than that of non-athletes at Harvard.

Athletes in sports such as American football, baseball and crew, however, tended to be more mesomorphic and endomorphic (Polednak & Damon, 1970; Polednak, 1972), and both somatotype components are positively associated with cardiovascular morbidity and mortality, and thus are negatively associated with longevity (Damon, 1965). Athletes in the major sports who won three or more varsity letters had a higher mortality from cardiovascular disease, specifically coronary heart disease, and were especially more mesomorphic than athletes in major sports who won only one or two varsity letters (Polednak, 1972). Thus, variation in physique may be associated with degree of success in a specific sport.

Most studies consider longevity of athletes who were active during the first half of this century; hence, the majority are probably white. Given the altered composition of intercollegiate teams over the past two decades, future studies will need to control for racial background. Additional concerns include major changes in training regimens, the use of performance enhancing substances, and extended schedules with frequent travel.

The available evidence for favourable longevity of former athletes is not conclusive. The association is complex. The highly select nature of athletic samples and factors such as degree of habitual physical activity after college and other indicators of life-style (smoking, weight gain and so on) need to be considered in evaluating the association between participation in intercollegiate sport and longevity. More recent analyses of data for Harvard alumni suggests that postcollegiate habitual physical activity and not past sport activity is associated with a low risk of coronary heart disease, while a sedentary life-style, even among ex-varsity athletes, is associated with a high risk of heart disease (Paffenbarger *et al.*, 1984).

In a more comprehensive analysis of the Harvard alumni data, Paffenbarger and colleagues (1986) estimated gains in longevity relative to habitual physical activity. Physically active alumni had significantly lower mortality rates and thus added years of life expectancy (Table 6.13). Physically active men (≥ 2000 kcal/week expended in walking, climbing

Table 6.13. *Physical activity, longevity and mortality in Harvard alumni**

Age at entry (years)	Estimated additional years of life (activity index†)		% surviving to age 80 (activity index†)	
	≥2000 vs <500	≥2000 vs <2000	≥2000	<2000
35–39	2.51	1.50	68.2	57.8
40–44	2.34	1.39	68.5	58.2
45–49	2.10	1.10	69.0	59.2
50–54	2.11	1.20	69.9	59.8
55–59	2.02	1.13	71.1	61.0
60–64	1.75	0.93	73.0	63.4
65–69	1.35	0.67	76.4	67.6
70–74	0.72	0.44	82.4	74.6
75–79	0.42	0.30	91.8	85.0

Adapted from Paffenbarger *et al.* (1986).
* All values are adjusted for differences in blood pressure status, cigarette smoking, net gain in the body mass index since college, and age of parental death.
† Kilocalories expended per week in walking, climbing stairs and playing sports.

stairs and playing sports) gained 2.15 years in longevity to the age of 80 than sedentary men (<500 kcal/week in physical activities). The gain in longevity is reduced to 1.25 years when the comparison is made between men having energy expenditures in physical activity ≥2000 kcal/week and <2000 kcal/week. When expressed in terms of relative survival, physically active men had an advantage of about 10 % over sedentary men from 35 to 64 years of age. These data which show a reasonably clear association between a physically active life-style and longevity are perhaps the most convincing to date.

References
AAHPER (1976). *AAHPER Youth Fitness Test Manual*. Reston, Va.: American Alliance for Health, Physical Education and Recreation.
AAHPERD (1980). *Health Related Physical Fitness Test Manual*. Reston, Va.: American Alliance for Health, Physical Education, Recreation and Dance.
   (1984). *Technical Manual Health Related Physical Fitness*. Reston, Va.: American Alliance for Health, Physical Education, Recreation and Dance.
Astrand, P. O., Engstrom, L., Eriksson, B. O., Karlberg, P., Nylander, I., Saltin, B. & Thoren, C. (1963). *Girl Swimmers. Acta Paediatrica Scandinavica, Supplement* **147**.
Ayers, J. W. T., Komesu, Y., Romain, T. & Ansbacher, R. (1985). Anthropometric, hormonal and psychologic correlates of semen quality in endurance trained male athletes. *Fertility and Sterility*, **43**, 917–21.

Bachmann, G. A. & Kemmann, E. (1982). Prevalence of oligomenorrhea and amenorrhea in a college population. *American Journal of Obstetrics and Gynecology*, **144**, 98–102.

Bailey, D. A., Malina, R. M. & Mirwald, R. L. (1986). Physical activity and growth of the child. In *Human Growth, Volume 2*, ed. F. Falkner & J. M. Tanner, pp. 147–70. New York: Plenum Press.

Baker, E. R. (1981). Menstrual dysfunction and hormonal status in athletic women: a review. *Fertility and Sterility*, **36**, 691–6.

Baker, E. R., Mathur, R. S., Kirk, R. F. & Williamson, H. O. (1981). Female runners and secondary amenorrhea: correlation with age, parity, mileage, and plasma hormonal and sex-hormone-binding globulin concentrations. *Fertility and Sterility*, **36**, 183–7.

Behrman, S. J. (1967). Adolescent amenorrhea. *Annals of the New York Academy of Sciences*, **142**, 807–12.

Berkowitz, G. S., Kelsey, J. L., Holford, T. R. & Berkowitz, R. L. (1983). Physical activity and the risk of spontaneous preterm delivery. *Journal of Reproductive Medicine*, **28**, 581–8.

Bielicki, T. & Charzewski, J. (1977). Sex differences in the magnitude of statural gains of offspring over parents. *Human Biology*, **49**, 265–77.

Blair, S. N. (1988). Exercise, health, and longevity. In *Perspectives in Exercise Science and Sports Medicine, Volume 1, Prolonged Exercise*, ed. D. R. Lamb and R. Murray, pp. 443–84. Indianapolis: Benchmark Press.

Blumenthal, J. A., O'Toole, L. C. & Chang, J. L. (1984). Is running an analogue of anorexia nervosa? *Journal of the American Medical Association*, **252**, 520–3.

Bonen, A., Belcastro, A. N., Simpson, A. A. & Ling, W. (1978). Comparison of LH and FSH concentrations in age group swimmers, moderately active girls, and adult women. In *Swimming Medicine IV*, ed. B. Eriksson & B. Furberg, pp. 70–8. Baltimore: University Park Press.

Borgen, J. S. & Corbin, C. B. (1987). Eating disorders among female athletes. *Physician and Sportsmedicine*, **15**, 89–95 (February).

Bouchard, C. (1986). Genetics of aerobic power and capacity. In *Sport and Human Genetics*, ed. R. M. Malina & C. Bouchard, pp. 59–88. Champaign, IL.: Human Kinetics.

Bouchard, C. & Malina, R. M. (1983). Genetics of physiological fitness and motor performance. *Exercise and Sport Sciences Reviews*, **11**, 306–39.

Brisson, G. R., Dulac, S., Peronnet, F. & Ledoux, M. (1982). The onset of menarche: a late event in pubertal progression to be affected by physical training. *Canadian Journal of Applied Sport Sciences*, **7**, 61–7.

Bullen, B. A., Skrinar, G. S., Beitins, I. Z., Carr, D. B., Reppert, S. M., Dotson, C. O., Fencl, M. de M., Gervino, E. V. & McArthur, J. W. (1984). Endurance training effects on plasma hormonal responsiveness and sex hormone excretion. *Journal of Applied Physiology: Respiratory, Environmental and Exercise Physiology*, **56**, 1453–63.

Bullen, B. A., Skrinar, G. S., Beitins, I. Z., von Mering, G., Turnbull, B. A. & McArthur, J. W. (1985). Induction of menstrual disorders by strenuous exercise in untrained women. *New England Journal of Medicine*, **312**, 1349–53.

Cadoux-Hudson, T. A., Few, J. D. & Imms, F. J. (1985). The effect of exercise on the production and clearance of testosterone in well trained young men. *European Journal of Applied Physiology*, **54**, 321–5.

CAHPER (1980). *CAHPER Fitness Performance II Test Manual*. Ottawa: Canadian Association for Health, Physical Education and Recreation.

Calabrese, L. H., Kirkendall, D. T., Floyd, M., Rapoport, S., Williams, G. W., Weiker, G. G. & Bergfeld, J. A. (1983). Menstrual abnormalities, nutritional patterns, and body composition in female classical ballet dancers. *Physician and Sportsmedicine*, **11**, 86–98 (February).

Carli, G., Martelli, G., Viti, A., Baldi, L., Bonifazi, M. & Lupo di Prisco, C. (1983a). The effect of swimming training on hormone levels in girls. *Journal of Sports Medicine and Physical Fitness*, **23**, 45–51.

——— (1983b). Modulation of hormone levels in male swimmers during training. In *Biomechanics and Medicine in Swimming*, ed. A. P. Hollander, P. A. Huijing & G. de Groot, pp. 33–40. Champaign, IL.: Human Kinetics.

Caspersen, C. J., Powell, K. E. & Christenson, G. M. (1985). Physical activity, exercise, and physical fitness: Definitions and distinctions for health-related research. *Public Health Reports*, **100**, 126–31.

Clapp, J. F. & Dickstein, S. (1984). Endurance exercise and pregnancy outcome. *Medicine and Science in Sports and Exercise*, **16**, 556–62.

Clarke, H. H. (1971). *Basic Understanding of Physical Fitness. Physical Fitness Research Digest, Series 1, no. 1*. Washington, DC: President's Council on Physical Fitness and Sports.

Collings, C. A., Curet, L. B. & Mullin, J. P. (1983). Maternal and fetal responses to a maternal aerobic exercise program. *American Journal of Obstetrics and Gynecology*, **145**, 702–7.

Cumming, D. C., & Rebar, R. W. (1983). Exercise and reproductive function in women. *American Journal of Industrial Medicine*, **4**, 113–25.

Dale, E., Gerlach, D. H. & Wilhite, A. L. (1979). Menstrual dysfunction in distance runners. *Obstetrics and Gynecology*, **54**, 47–53.

Dale, E., Mullinax, K. M. & Bryan, D. H. (1982). Exercise during pregnancy: effects on the fetus. *Canadian Journal of Applied Sport Science*, **7**, 98–103.

Dalton, K. (1968). Menstruation and examinations. *Lancet*, **2**, 1386–8.

Damon, A. (1965). Delineation of the body build variables associated with cardiovascular diseases. *Annals of the New York Academy of Sciences*, **126**, 711–27.

Dann, T. C. & Roberts, D. F. (1984). Menarcheal age in University of Warwick students. *Journal of Biosocial Science*, **16**, 511–19.

Davison, A. E. (1981). 'The age at menarche and selected familial and menstrual characteristics of high school varsity athletes.' Master's thesis, University of Texas at Austin.

Dessypris, A., Kuoppasalmi, K. & Adlercreutz, H. (1976). Plasma cortisol, testosterone, androstenedione and luteinising hormone (LH) in a noncompetitive marathon run. *Journal of Steroid Biochemistry*, **7**, 33–7.

Devor, E. J. & Crawford, M. H. (1984). Family resemblance for neuromuscular performance in a Kansas Mennonite community. *American Journal of Physical Anthropology*, **64**, 289–96.

Dobzhansky, Th. (1962). *Mankind Evolving: the evolution of the human species.* New Haven: Yale University Press.

Drinkwater, B. L. (1984). Athletic amenorrhea: a review. In *Exercise and Health, American Academy of Physical Education Papers No. 17*, pp. 120–31. Champaign, IL.: Human Kinetics.

Drinkwater, B. L., Nilson, K., Chesnut, C. H., Bremner, W. J., Shainholtz, S. & Southworth, M. B. (1984). Bone mineral content of amenorrheic and eumenorrheic athletes. *New England Journal of Medicine*, **311**, 277–81.

Ellison, P. T. & Lager, C. (1986). Moderate recreational running is associated with lowered salivary progesterone profiles in women. *American Journal of Obstetrics and Gynecology*, **154**, 1000–3.

Ellison, P. T., Lager, C. & Calfee, J. (1987). Low profiles of salivary progesterone among college undergraduate women. *Journal of Adolescent Health Care*, **8**, 204–7.

Erdelyi, G. J. (1962). Gynecological survey of female athletes. *Journal of Sports Medicine and Physical Fitness*, **2**, 174–9.

Eriksson, B. O., Engstrom, I., Karlberg, P., Lundin, A., Saltin, B. & Thoren, C. (1978). Long-term effect of previous swimtraining in girls. A 10-year follow-up of the 'Girl Swimmers'. *Acta Paediatrica Scandinavica*, **67**, 285–92.

Erkkola, R. (1976). The influence of physical training during pregnancy on physical work capacity and circulatory parameters. *Scandinavian Journal of Clinical and Laboratory Investigation*, **36**, 747–54.

Erkkola, R. & Makela, M. (1976). Heart volume and physical fitness of parturients. *Annals of Clinical Research*, **8**, 15–21.

Fahey, T. D., del Valle-Zuris, A., Oehlsen, G., Trieb, M. & Seymour, J. (1979). Pubertal stage differences in hormonal and hematological responses to maximal exercise in males. *Journal of Applied Physiology*, **46**, 823–7.

Fraioli, F., Cappa, M., Fabbri, A., Gnessi, L., Moretti, C., Borrelli, P. & Isidori, A. (1984). Lack of endogenous opioid inhibitory tone on LH secretion in early puberty. *Clinical Endocrinology*, **20**, 299–305.

Frisch, R. E. (1976). Fatness of girls from menarche to age 18 years, with a nomogram. *Human Biology*, **48**, 353–9.

(1981). Athletics can combat teen-age pregnancy. *New York Times*, 16 July (letter).

(1988). Fatness and fertility. *Scientific American*, **258**, 88–95 (March).

Frisch, R. E., Gotz-Welbergen, A. B., McArthur, J. W., Albright, T., Witschi, J., Bullen, B., Birnholz, J., Reed, R. B. & Hermann, H. (1981). Delayed menarche and amenorrhea of college athletes in relation to age of onset of training. *Journal of the American Medical Association*, **246**, 1559–63.

Galbo, H., Hummer, L., Petersen, I. B., Christensen, N. J. & Bie, N. (1977). Thyroid and testicular hormone responses to graded and prolonged exercise in man. *European Journal of Applied Physiology*, **36**, 101–6.

Garn, S. M., LaVelle, M. & Pilkington, J. J. (1983). Comparisons of fatness in premenarcheal and postmenarcheal girls of the same age. *Journal of Pediatrics*, **103**, 328–31.

Garrison, R. J., Anderson, B. E. & Reed, S. C. (1968). Assortative marriage. *Eugenics Quarterly*, **15**, 113–27.

Gawel, M. J., Park, D. M., Alaghband-Zadeh, J. & Rose, C. F. (1979). Exercise and hormonal secretion. *Postgraduate Medical Journal*, **55**, 373–6.

Gorski, J. (1985). Exercise during pregnancy: maternal and fetal responses. A brief review. *Medicine and Science in Sports and Exercise*, **17**, 407–16.

Greene, B. B., Daling, J. R., Weiss, N. S., Liff, J. M. & Koepsell, T. (1986). Excercise as a risk for infertility with ovulatory dysfunction. *American Journal of Public Health*, **76**, 1432–6.

Guglielmini, C., Paolini, A. R. & Conconi, F. (1984). Variations of serum testosterone concentrations after physical exercises of different duration. *International Journal of Sports Medicine*, **5**, 246–9.

Hamilton, L. H., Brooks-Gunn, J., Warren, M. P. & Hamilton, W. G. (1988). The role of selectivity in the pathogenesis of eating problems in ballet dancers. *Medicine and Science in Sports and Exercise*, **20**, 560–5.

Harrison, G. A., Givson, J. B. & Hiorns, R. W. (1976). Assortative marriage for psychometric, personality, and anthropometric variation in a group of Oxfordshire villages. *Journal of Biosocial Science*, **8**, 145–53.

Jaszmann, L., Van Lith, N. D. & Zaat, J. C. A. (1969). The age of menopause in the Netherlands. *International Journal of Fertility*, **14**, 106–17.

Johnson, R. C., DeFries, J. C., Wilson, L. R., McClearn, G. E., Vandenberg, S. G., Ashton, G. C., Mi, M. P. & Rashad, M. N. (1976). Assortative marriage for specific cognitive abilities in two ethnic groups. *Human Biology*, **48**, 343–52.

Johnston, F. E., Roche, A. F., Schell, L. M. & Wettenhall, N. B. (1975). Critical weight at menarche: a critique of a hypothesis. *American Journal of Diseases of Children*, **129**, 19–23.

Journal of Physical Education and Recreation (1979). Definition of physical fitness. *Journal of Physical Education and Recreation*, **50**, 28 (October).

Kaiserauer, S., Snyder, A. C., Sleeper, M. & Zierath, J. (1989). Nutritional, physiological, and menstrual status of distance runners. *Medicine and Science in Sports and Exercise*, **21**, 120–5.

Khoury, S. A., Reame, N. E., Kelch, R. P. & Marshall, J. C. (1987). Diurnal patterns of pulsatile luteinizing hormone secretion in hypothalamic amenorrhea: reproducibility and responses to opiate blockade and an α_e-adrenergic agonist. *Journal of Clinical Endocrinology and Metabolism*, **64**, 755–62.

Kovar, R. (1981). *Human Variation in Motor Abilities and its Genetic Analysis*. Prague: Charles University.

Lesage, R., Simoneau, J.-A., Jobin, J., Leblanc, C. & Bouchard, C. (1985). Familial resemblance in maximal heart rate, blood lactate and aerobic power. *Human Heredity*, **35**, 182–9.

Liestol, K. (1980). Menarcheal age and spontaneous abortion: a causal connection? *American Journal of Epidemiology*, **111**, 753–8.

Lortie, G., Bouchard, C., Leblanc, C., Tremblay, A., Simoneau, J.-A., Theriault, G. & Savoie, J. P. (1982). Familial similarity in aerobic power. *Human Biology*, **54**, 801–12.

Lotgering, F. K., Gilbert, R. D. & Longo, L. D. (1985). Maternal and fetal responses to exercise during pregnancy. *Physiological Reviews*, **65**, 1–36.

Loucks, A. B. & Horvath, S. M. (1985). Athletic amenorrhea: a review. *Medicine and Science in Sports and Exercise*, **17**, 56–72.

McArthur, J. W., Bullen, B. A., Beitins, I. Z., Pagano, M., Badger, T. M. & Klibanski, A. (1980). Hypothalamic amenorrhea in runners of normal body composition. *Endocrine Research Communications*, **7**, 13–25.

McColl, E. M., Wheeler, G. D., Gomes, P., Bhambhani, Y. & Cumming, D. C. (1989). The effects of acute exercise on pulsatile LH release in high-mileage male runners. *Clinical Endocrinology*, **31**, 617–21.

MacConnie, S. E., Barkan, A., Lampman, R. M., Schork, M. A. & Beitins, I. Z. (1986). Decreased hypothalamic gonadotropin-releasing hormone secretion in male marathon runners. *New England Journal of Medicine*, **315**, 411–17.

Malina, R. M. (1978). Adolescent growth and maturation: selected aspects of current research. *Yearbook of Physical Anthropology*, **21**, 63–94.

(1983a). Menarche in athletes: a synthesis and hypothesis. *Annals of Human Biology*, **10**, 1–24.

(1983b). Human growth, maturation and regular physical activity. *Acta Medica Auxologica*, **15**, 5–27.

(1986). Genetics of motor development and performance. In *Sport and Human Genetics*, ed. R. M. Malina & C. Bouchard, pp. 23–58. Champaign, IL.: Human Kinetics.

(1988). Biological maturity status of young athletes. In *Young Athletes: biological, psychological, and education perspectives*, ed. R. M. Malina, pp. 121–40. Champaign, IL.: Human Kinetics.

Marker, K. (1983). *Frau und Sport*. Leipzig: Johann Ambrosius Barth.

Martin, E. J., Brinton, L. A. & Hoover, R. (1983). Menarcheal age and miscarriage. *American Journal of Epidemiology*, **117**, 634–6.

Mauras, N., Veldhuis, J. D. & Rogol, A. D. (1986). Role of endogenous opiates in pubertal maturation: opposing actions of naltrexone in prepubertal and late pubertal boys. *Journal of Clinical Endocrinology and Metabolism*, **62**, 1256–63.

Montoye, H. J. & Gayle, R. (1978). Familial relationships in maximal oxygen uptake. *Human Biology*, **50**, 241–9.

National Center for Health Statistics (1977). Teenage childbearing: United States, 1966–1975. *Monthly Vital Statistics Report*, **26**, no. 5.

Niemineva, K. (1953). On the course of delivery of Finnish baseball (Pesapallo) players and swimmers. In *Sport Medicine*, ed. M. J. Karvonen, pp. 169–172. Helsinki: Finnish Association of Sports Medicine.

Oswald, J. (1980). Sleep as a restorative process: human clues. In *Adaptive Capabilities of the Nervous System*, ed. P. S. McConnell, G. J. Boer, H. J. Romijn, N. E. van de Poll & M. A. Corner, pp. 279–88. New York: Elsevier.

Paffenbarger, R. S., Hyde, R. T., Wing, A. L. & Hsieh, C.-C. (1986). Physical activity, all-cause mortality, and longevity of college alumni. *New England Journal of Medicine*, **314**, 605–13.

Paffenbarger, R. S., Hyde, R. T., Wing, A. L. & Steinmetz, C. H. (1984). A natural history of athleticism and cardiovascular health. *Journal of the American Medical Association*, **252**, 491–5.

Peltenburg, A. L., Erich, W. B. M., Thijssen, J. J. H., Veeman, W., Jansen, M., Bernink, M. J. E., Zonderland, M. L., van den Brande, J. L. & Huisveld, I.

182   R. M. Malina

A. (1984). Sex hormone profiles of premenarcheal athletes. *European Journal of Applied Physiology*, **52**, 385–92.

Polednak, A. P. (1972). Longevity and cardiovascular mortality among former college athletes. *Circulation*, **46**, 649–54.

Polednak, A. P. & Damon, A. (1970). College athletics, longevity, and cause of death. *Human Biology*, **42**, 28–46.

Pomerance, J. J., Gluck, L. & Lynch, V. A. (1974). Physical fitness in pregnancy: its effect on pregnancy outcome. *American Journal of Obstetrics and Gynecology*, **119**, 867–76.

Powell, K. E., Thompson, P. D., Caspersen, C. J. & Kendrick, J. S. (1987). Physical activity and the incidence of coronary heart disease. *Annual Review of Public Health*, **8**, 253–87.

Prior, J. C. (1982). Endocrine 'conditioning' with endurance training: a preliminary review. *Canadian Journal of Applied Sports Sciences*, **7**, 149–57.

Prior, J. C., Ho Yuen, B., Clement, P., Bowie, L. & Thomas, J. (1982). Reversible luteal phase changes and infertility associated with marathon training. *Lancet*, **2**, 269–70.

Reeves, J. (1979). Estimating fatness. *Science*, **204**, 881.

Reiff, G. G., Dixon, W. R., Jacoby, D., Ye, G. X., Spain, C. G. & Hunsicker, P. A. (1986). *The President's Council on Physical Fitness and Sports 1985 National School Population Fitness Survey*. Ann Arbor: University of Michigan.

Roberts, D. F. (1977). *Assortative Mating in Man: husband/wife correlations in physical characteristics*. *Bulletin of the Eugenics Society, Supplement* **2**.

Roberts, D. F. & Dann, T. C. (1967). Influences on menarcheal age in girls in a Welsh college. *British Journal of Preventive and Social Medicine*, **21**, 170–6.

Roberts, D. F., Danskin, M. J. & Chinn, S. (1975). Menarcheal age in Northumberland. *Acta Paediatrica Scandinavica*, **64**, 845–52.

Roberts, D. F., Rozner, L. M. & Swan, A. V. (1971). Age at menarche, physique and environments in industrial North East England. *Acta Paediatrica Scandinavica*, **60**, 158–64.

Rogol, A. D., Weltman, J., Seip, R. L., Veldhuis, J. D., Evans, S. & Weltman, A. (1989). Gonadotropin secretion in female runners. In *Hormones and Sport*, ed. Z. Laron & A. D. Rogol, pp. 141–66. New York: Raven Press.

Rosen, L. W., McKeag, D. B., Hough, D. O. & Curley, V. (1986). Pathogenic weight-control behavior in female athletes. *Physician and Sportsmedicine*, **14**, 79–86 (January).

Sandler, D. P., Wilcox, A. J. & Horney, L. F. (1984). Age at menarche and subsequent reproductive events. *American Journal of Epidemiology*, **119**, 765–74.

Schmitt, W. M., Schnabel, A. & Kindermann, W. (1983). Hormonal responses to marathon running in children and adolescents. *International Journal of Sports Medicine*, **4**, 68 (abstract).

Schwartz, B., Cumming, D. C., Riordan, E., Selye, M., Yen, S. S. C. & Rebar, R. W. (1981). Exercise-associated amenorrhea: a distinct entity? *American Journal of Obstetrics and Gynecology*, **141**, 662–70.

Scott, E. C. & Johnston, F. E. (1982). Critical fat, menarche, and the maintenance of menstrual cycles: a critical review. *Journal of Adolescent Health Care*, **2**, 249–60.

(1985). Science, nutrition, fat, and policy: tests of the critical-fat hypothesis. *Current Anthropology*, **26**, 463–73.

Seefeldt, V., Haubenstricker, J., Branta, C. & Evans, S. (1988). Physical characteristics of elite young distance runners. In *Competitive Sports for Children and Youth*, ed. E. W. Brown & C. F. Branta, pp. 247–58. Champaign IL.: Human Kinetics.

Shangold, M. M. (1984). Exercise and the adult female: Hormonal and endocrine effects. *Exercise and Sport Sciences Reviews*, **12**, 53–79.

(1988). Menstruation. In *Women and Exercise: physiology and sports medicine*, ed. M. M. Shangold & G. Mirkin, pp. 129–44. Philadelphia: F. A. Davis.

Shangold, M. M., Freeman, R., Thysen, B. & Gatz, M. (1979). The relationship between long-distance running, plasma progesterone, and luteal phase length. *Fertility and Sterility*, **31**, 130–3.

Shangold, M. M. & Levine, H. S. (1982). The effect of marathon training upon menstrual function. *American Journal of Obstetrics and Gynecology*, **143**, 862–9.

Simons, J. & Renson, R. (eds) (1982). *Evaluation of Motor Fitness*. Leuven, Belgium: Institute of Physical Education, Catholic University of Leuven.

Speroff, L. & Redwine, D. B. (1980). Exercise and menstrual function. *Physician and Sportsmedicine*, **8**, 42–52.

Spuhler, J. N. (1968). Assortative mating with respect to physical characteristics. *Eugenics Quarterly*, **15**, 128–40.

Stager, J. M., Robsertshaw, D. & Miescher, E. (1984). Delayed menarche in swimmers in relation to age at onset of training and athletic performance. *Medicine and Science in Sports and Exercise*, **16**, 550–5.

Stager, J. M., Wigglesworth, J. K. & Hatler, L. K. (1990). Interpreting the relationship between age of menarche and prepubertal training. *Medicine and Science in Sports and Exercise*, **22**, 54–8.

Stephens, K. E., Van Huss, W. D., Olson, H. W. & Montoye, H. J. (1984). The longevity, morbidity, and physical fitness of former athletes – an update. In *Exercise and Health*, American Academy of Physical Education Papers No. 17, pp. 101–19. Champaign, IL.: Human Kinetics.

Stinson, S. (1985). Sex differences in environmental sensitivity during growth and development. *Yearbook of Physical Anthropology*, **28**, 123–47.

Strauss, R. H., Lanese, R. R. & Malarkey, W. B. (1985). Weight loss in amateur wrestlers and its effect on serum testosterone levels. *Journal of the American Medical Association*, **254**, 3337–8.

Stukovsky, R., Valsik, J. A. & Bulai-Stirbu, M. (1967). Family size and menarcheal age in Constanza, Roumania. *Human Biology*, **39**, 277–83.

Susanne, C. (1967). Contribution à l'étude de l'assortiment matrimonial dans un échantillon de la population Belge. *Bulletin de la Société Royale Belge d'Anthropologie et de Préhistoire*, **78**, 147–96.

Sutton, J. R., Coleman, M. J., Casey, J. & Lazarus, L. (1973). Androgen responses during physical exercise. *British Medical Journal*, **1**, 520–2.

184   R. M. Malina

Szopa, J. (1982). Familial studies on genetic determination of some manifes-
tations of muscular strength in man. *Genetica Polonica*, **23**, 65–79.
Tanner, J. M. (1962). *Growth at Adolescence* (2nd edn). Oxford: Blackwell
Scientific Publications.
Treloar, A. E. (1974). Menarche, menopause, and intervening fecundity. *Human
Biology*, **46**, 89–107.
Trussell, J. (1980). Statistical flaws in evidence for the Frisch hypothesis that
fatness triggers menarche. *Human Biology*, **52**, 711–20.
Veldhuis, J. D., Evans, W. S., Demers, L. M., Thorner, M. O., Wakat, D. &
Rogol, A. D. (1985). Altered neuroendocrine regulation of gonadotropin
secretion in women distance runners. *Journal of Clinical Endocrinology and
Metabolism*, **61**, 557–63.
Wall, S. R. & Cumming, D. C. (1985). Effects of physical activity on reproductive
function and development in males. *Seminars in Reproductive Endocrinology*, **3**, 65–80.
Warren, M. P. (1980). The effects of exercise on pubertal progression and
reproductive function in girls. *Journal of Clinical Endocrinology and Metabolism*, **51**, 1150–7.
(1983). Effects of undernutrition on reproductive function in the human.
Endocrine Reviews, **4**, 363–77.
Warren, M. P. & Brooks-Gunn, J. (1989). Delayed menarche in athletes: The
role of low energy intake and eating disorders and their relation to bone
density. In *Hormones and Sport*, ed. Z. Laron & A. D. Rogol, pp. 41–54.
New York: Raven Press.
Weitzman, E. D. (1980). Biological rhythms and hormone secretion patterns. In
Neuroendocrinology, ed. D. T. Krieger & J. C. Hughes, pp. 85–92. Sunder-
land, Massachusetts: Sinauer Associates.
Wellens, R. E. (1984). 'The influence of sociocultural variables and sports
participation on the age at menarche of Flemish girls (The Leuven Growth
Study of Flemish Girls).' Master's thesis, University of Texas at Austin.
Wheeler, G. D., Wall, S. R., Belcastro, A. N. & Cumming, D. C. (1984).
Reduced serum testosterone and prolactin levels in male distance runners.
Journal of the American Medical Association, **252**, 514–16.
Wierman, M. E. & Crowley, W. F. Jr (1986). Neuroendocrine control of the
onset of puberty. In *Human Growth. Volume 2, Postnatal Growth, Neuro-
biology*, ed. F. Falkner & J. M. Tanner, pp. 225–41. New York: Plenum
Press.
Wolanski, N. (1973). Assortative mating in the rural Polish populations. *Studies
in Human Ecology*, **1**, 182–8.
Yates, A., Leehey, K., & Shisslak, C. (1983). Running: an analogue of anorexia?
New England Journal of Medicine, **308**, 251–5.
Zaharieva, E. (1972). Olympic participation by women: effects on pregnancy and
childbirth. *Journal of the American Medical Association*, **221**, 992–5

7 Human evolution and the genetic epidemiology of chronic degenerative diseases

DOUGLAS E. CREWS AND GARY D. JAMES

Degenerative diseases have likely plagued *Homo sapiens* for most of the species' evolutionary history (Hinkle, 1987). Their prevalence now is clearly far greater than at any previous time. One reason for this increase may be related to the fact that degenerative disease generally afflicts people over 40, a segment of human populations which has disproportionately grown in Western industrialized populations as survivorship has increased to six decades and longer (Adams & Smouse, 1985; Stini, this volume).

Finding the genetic basis for diseases such as hypertension, coronary heart disease (CHD), cancer, and non-insulin-dependent diabetes mellitus (NIDDM) has proved to be a formidable challenge. One of the main reasons for the difficulty is that many chronic diseases are defined (in general) as a pathological endpoint of several polygenic traits. The heritability of these diseases is thus related to the heritability of the traits for which they represent pathology, such as blood pressure, serum cholesterol, plasma glucose levels and body fat. In addition, because of the polygenic nature of these traits, environment significantly contributes to their phenotypic expression and hence also plays a major role in the development of the disease state. This discussion will focus on the conceptual and methodological problems inherent in studying the genetics of chronic disease, and how the perspective of human population biology and biological anthropology may improve our understanding of why these diseases arose and how they may be inherited.

Human evolution and chronic diseases

It is probably incorrect to assume that chronic diseases are only a manifestation of our own modern life-styles, since whatever genes are responsible for them have likely been with us since even before the appearance of our modern form some 35 000 to 40 000 years ago (Hinkle,

1987; Easton, Konner & Shostak, 1988). There is excellent evidence that our ancesteral populations were not immune. For example, atherosclerosis was present in Egypt some 5000 years ago (Hinkle, 1987) and one of the first human fossils found (Neanderthal) suffered from osteoarthritis (Williams, 1973).

The reasons behind the present-day epidemic of chronic disease are many. Degenerative diseases occur most frequently in people who survive past the fourth decade of life. This segment of human populations was very small in even European populations only 300 years ago (Hinkle, 1987). Given the small group of persons at risk and the few cases that met medical attention (such as it was), it is not surprising that the medical definition and diagnosis of many of these diseases has waited for the present century (Hinkle, 1987). Thus, our present 'epidemic' may be partly explained by the greater number of people living well beyond 40 in Western industrialized nations and improvements in medical diagnosis. However, these demographic and medical facts provide few clues as to how these diseases evolved, why they perpetuate, and why they seem only to affect individuals aged 40 years and older.

Traits which affect mortality and reproduction either before or during the reproductive years have a significant natural selection against them (Williams, 1957). Thus, genes which dictated the development of NIDDM, cardiovascular diseases (CVD), or cancer at relatively young ages were likely selected out of early *Homo* populations, leaving genes that either precluded individuals from developing chronic diseases at any time, or dictated their development at a later stage in life (that is, beyond the lifespan necessary to attain reproductive success at that human evolutionary horizon). It is also possible that the genes which cause these conditions conferred a selective advantage in some individuals at a young age or during reproduction; thus there may also have been a positive selection for late-acting degenerative disease genes (Williams, 1957).

However, for the last 10 000 years, and particularly over the past few decades, the human environment has also changed so dramatically (Harrison, 1973; Baker & Baker, 1977; Baker, 1984) that genes which may have conferred selective advantage in our hunting and gathering past may put us at a selective disadvantage today (Levi & Anderson, 1975; Eaton *et al.*, 1988; James, Crews & Pearson, 1989). It is not unreasonable to suggest therefore, that another contributor to the increasing prevalence of degenerative diseases in Western populations today is the lack of fit between our ancient genes and our present-day environment (James & Baker, 1990). While a particular focus has been on the change in the dietary environment (Stini, 1971; Eaton & Konner, 1985), the social and physical environments are also drastically changed (and changing) (Baker, 1984). All of these alterations are significantly affecting our evolution, principally through their impact on the development of degenerative diseases. Thus, in addition to demographics, there are both significant genetic and significant environmental reasons for the presence and persistence of degenerative diseases in modern populations.

Degenerative diseases: definitions and diagnoses

Medically, chronic disease states are often diagnosed by a specific cut-off point of a continuously distributed metabolic or physiologic trait (Sing, Boerwinkle & Moll, 1985). When the value of some measurement exceeds the cut-off, disease is considered present. These cut-offs, such as that with blood pressure (in determining hypertension), are in many ways arbitrary (Pickering, 1961), although they are often based on retrospective or prospective population data which show that individuals with elevated or depressed values are at greater risk of dying. The critical value of disease is set at some inflection point in the survivorship curve. Interestingly, as time progresses, the value of the disease cut-off changes (such as that with cholesterol or blood pressure) (AHA, 1988). Another point about metabolic and physiological traits in the chronic disease process is that they are seasonally and diurnally quite variable within the individual, so that it may also be relevant to ask how long (or often) out of a 24-hour period values must be excessively out of a defined 'normal' range for the individual to be considered diseased (Pickering *et al.*, 1985). The only certain way of ascertaining that cardiovascular disease, for example, is present, is after a stroke or myocardial infarction has occurred (that is, after a life-threatening breakdown of the diseased system).

Another problem in defining chronic diseases is that many of them occur simultaneously in a kind of syndrome (for instance, NIDDM with hypertension and renal failure) (see, for example, Laragh & Brenner, 1990). It is often difficult to ascribe any specific pathology to a single specific chronic disease state. Genetically, this may suggest that a single gene has effects in several metabolic systems, thus causing multiple 'diseases'.

A final point is that diseases such as cancer are genetically the result of non-heritable somatic mutations. If these diseases have a heritable component, it is in the inheritance of specific genes (oncogenes) which are 'fragile', in the sense that they have a greater propensity to mutate (Yunis & Hoffman, 1989).

Risk factors and genetics of chronic diseases

The epidemiological and clinical focus on risk factors for defined morbid and mortal events that result from chronic diseases has, in many ways, hampered the search for their genetic basis. Because chronic diseases are polygenic in origin, variants at any one of several loci may result in the same disease phenotype (Sing et al., 1985). Lumping the phenotypes together and then statistically examining demographic, psychological and physiological associations may obfuscate the true genetic and/or environmental causes of the disease state (James & Baker, 1990). Since probands are defined medically by their phenotypes, actual gene–environment interactions may be under- or over-estimated or even missed, given the traditional linear modelling techniques of standard epidemiology.

Genetic epidemiological approaches

There are a limited number of approaches that can be used to assess the genetic basis of chronic diseases (Sing et al., 1985; Ward, 1985, 1990). One set of methods relies on the initial diagnosis of clinical disease in families or population segments. Pedigrees or the DNA of specific probands are examined for possible gene markers or variants at specific loci that define disease. Often, analysis is limited to only those diagnosed as diseased, so that the genetic variance within the phenotype can be characterized. Conceptually, this kind of approach can be thought of as redefining the chronic disease out of existence. That is, there is an attempt to identify major loci or the specific underlying genotypes, so that the disease phenotype may be thought of as a common symptom of several inborn errors of metabolism.

Another approach is one which employs classical quantitative genetics as described by Falconer (1960) in which the phenotypic variance in the population is partitioned into its genetic and environmental components. In this approach, there is a greater emphasis on defining the environmental components of the diseases (Ward, 1985, 1990) and the medical 'cut-off' point of disease is ignored (that is, the total healthy and diseased population is pooled).

Both of these approaches have been useful in disentangling the complexity of the genetic basis of chronic disease. In the following sections we review some of what has been found.

Blood pressure

Blood pressure, and hypertension in particular, is a complex trait in which the clinically defined phenotype may arise through a wide array of pathophysiological mechanisms (Ward, 1990). Several family environmental variables have been identified that are associated with blood pressure. These include excessive weight/obesity, sodium intake, and levels of physical activity, characteristics that tend to aggregate in both hypertensive and non-hypertensive families (Ward, 1990). The single most important of these for blood pressure is relative weight/obesity. A major problem in determining the genetic heritability of blood pressure is confounding by these environmentally labile characteristics.

Genetic heritability of blood pressure, estimated from a number of samples, ranges from 0.25 to 0.41 for systolic and from 0.18 to 0.35 for diastolic blood pressure (Glueck *et al.*, 1985; see also Ward, 1990). Although genetic factors have been shown to influence blood pressure, specific metabolic traits that lead to elevated blood pressure have not as yet been established. One candidate for this distinction is Na–Li countertransport; variation at this locus has been shown to influence blood pressure and may account for a substantial proportion of systolic hypertension (Boerwinkle *et al.*, 1986; Ward, 1990). Examination of various genetic models suggests that polygenic factors influenced by environment better fit available hypertension data than do models of a single or several major genes segregating in human populations (Pickering, 1967; Morton *et al.*, 1980; Ward, 1990).

Dietary Na may be associated with high blood pressure and hypertension and thus secondarily influence cardiovascular related morbidity and mortality (INTERSALT, 1988). Within populations sodium excretion apparently is significantly correlated with blood pressure. These associations also have been reported in cross-cultural analyses, although they are statistically weak (INTERSALT, 1988).

Sodium availability is low in inland tropical areas such as central Africa and South America (Wilson, 1986; INTERSALT, 1988). Populations that reside in these ecological settings may show higher sodium retention than comparable populations from salt- or sodium-rich areas. A genetic predisposition for enhanced retention of sodium might then be correlated with high blood pressure and fluid volume in a high-sodium environment. No association with blood pressure might be observed during youth, because reserve capacity to handle excess salt or fluid would still be present and age-related functional declines would not have commenced. However, susceptible genotypes might show greater fluid retention and consequent earlier physiological change in blood pressure as they aged. Genes associated with Na retention or transport may explain some differences in blood pressure and prevalences of hypertension of American blacks and whites, and place of origin in Africa may explain some part of the discordant results in different samples of US blacks.

Physiological mechanisms that produce a statistical association between adiposity/obesity/body weight and blood pressure have not been studied. However, they may involve differences in sodium excretion, storage, or retention related to obesity. Hyperinsulinemia is a secondary effect of obesity, and obesity is associated with both NIDDM and hypertension (McGarvey et al., 1989). Hyperinsulinemia leads to increased tubular resorption of Na by the kidneys, thus hyperinsulinemia may be the missing link between obesity and hypertension (Christlieb et al., 1985; Lucas et al., 1985).

Cholesterol and apolipoproteins

Cholesterol is endogenous in tissues that produce large amounts of steroids. These sources apparently provide adequate amounts of cholesterol for adult needs. During pregnancy, serum cholesterol rises and remains high for several months, providing additional sources for the growing and developing foetus. Plasma total cholesterol and LDL-cholesterol are influenced by dietary cholesterol. About 40 % of ingested cholesterol is absorbed, the remainder is passed in the stool (Connor & Connor, 1985). Excretion pathways are limited and most circulating cholesterol is returned to the liver, where it is recirculated (Conner & Connor, 1985). Efficient reabsorption may be secondary to genetically determined mechanisms and suggests that retention of cholesterol was probably necessary during previous phases of human evolution.

In the population of North America, genetic disorders of lipid metabolism are observed in about 2–3 % of the population, and constitute the most common genetic abnormality seen in primary practices (Illingworth & Connor, 1985). Both hyper- and hypo-lipidemia syndromes and various associated apolipoprotein disorders are recognized, but the precise genetic basis for these traits is not well documented. Much of the world population exhibits cholesterol- or lipid-conserving mechanisms, presumably of genetic origin. Familial hypercholesterolemia, familial hypertriglyceridemia, and overproduction of very low-density lipoprotein (VLDL), and other rare forms of hyperlipidemia may fall into this category (Glueck et al., 1985). Although uncommon in Euro-American samples, such disorders may affect as many as 5% of individuals worldwide (Glueck *et al.*, 1985). However, the prevalence of such familial syndromes in more traditional-living populations is generally undetermined.

Data from the Lipid Research Clinics Family Study indicate major gene(s) involved in hyperalphalipoproteinemia (high HDL-cholesterol), hyperbetalipoproteinemia (high LDL-cholesterol), and hypobetalipoproteinemia (low LDL-cholesterol), although results could also be interpreted as indicating discrete environmental factors transmitted from parents to some children but not others (Green *et al.*, 1984). HDL-cholesterol is inversely correlated with risk of CHD and differences between individuals may be related to genetic variation in apolipoproteins or lipid receptors. One fundamental metabolic difference in all populations examined is that men have lower HDL-cholesterol than women (Illingworth & Connor, 1985).

One genetic factor closely related to cholesterol metabolism is LDL-receptor (LDL-R). A large number of LDL-Rs is associated with lower cholesterol and higher plasma HDL-cholesterol. The number and efficiency of LDL-Rs is genetically determined (Brown & Goldstein, 1989). LDL-Rs are important determinants of plasma LDL levels. The number of available LDL-Rs is genetically and environmentally determined by a biofeedback system that suppresses manufacture of LDL-R, by up to 90%, in the presence of high plasma LDL (Brown & Goldstein, 1989). Interestingly, Pima Indians, a relatively obese population with a high-fat diet, show high numbers of LDL-Rs, low plasma cholesterol, and low LDL-cholesterol levels (Brown & Goldstein, 1989). This suggests that positive genetic factors involved in lipid metabolism may be segregating in Pima Indians.

Genetic variation at various apolipoprotein (APO) loci also affects plasma lipid levels (Davignon, Gregg & Sing, 1988; Ferrell, 1989). Variation in APO E is associated with severe hypertriglyceridemia and APO B with familial hypertriglyceridemia (Illingworth & Connor, 1985). Variation in three alleles at the APO E locus may explain up to 7% of total variation in plasma cholesterol in Caucasian populations, with APO E-II contributing a cholesterol-lowering effect and APO E-IV a cholesterol-raising effect (Davignon *et al.*, 1988).

Spouse–spouse correlations for lipids and lipoproteins are small and, since longer-married couples show higher correlations, are probably due to shared environments. Parent–offspring correlations for cholesterol, HDL- and LDL-cholesterol are generally significant for non-adult children, and persist at lower levels for adult offspring not sharing a parental household. In addition, heritability estimates for all lipid measurements are commonly greater than those for systolic and diastolic blood pressure or adiposity (Glueck *et al.*, 1985).

Cholesterol levels are low among traditional agricultural and foraging populations compared with more cosmopolitan societies (see, for example, Connor *et al.*, 1978; Mancilha-Carvalho & Crews, 1990). This disparity has been attributed to ecological factors related to nutrient availability in high-fibre, low-fat, low-caloric, and low-cholesterol traditional and wild foods compared to the low-fibre, high-fat, high-caloric, and cholesterol-dense diets consumed in cosmopolitan societies (Trowel & Burkitt, 1981; Eaton *et al.*, 1988). In traditional societies cholesterol-conserving mechanisms might be advantageous. However, specific genetic markers for familial hypercholesterolemia and hypertriglyceridemia are needed to examine the distribution and survival/reproductive consequences of these traits in low-cholesterol environments. Gross lipid differences may not occur, even among individuals with these traits, in the absence of high levels of dietary cholesterol and/or fats. Polynesian populations of the South Pacific may be of interest with regard to dietary influences on lipid levels. These groups rely on the coconut as a major food source. Coconuts are a calorie-dense food with a large proportion of calories in the form of saturated fats. Populations that have subsisted for long periods of time in an ecological setting with high quantities of such readily available dietary fats or cholesterol should have fewer genes associated with cholesterol conservation and retention. This may be true for Samoans, a group noted for obesity and highly prevalent NIDDM. At all ages, they show average plasma total cholesterol lower and HDL-cholesterol absolutely lower, but relatively greater, than that of United States samples of much lower body weight (Pelletier & Hornick, 1986). Prevalence estimates for familial hypercholesterolemia and hypertriglyceridemia are not available for Samoans. Thus, the available data suggest, but do not show, that prevalences will be below those in populations where dietary fats and cholesterol have been scarce until recent centuries.

Fats, adipose tissue, fat patterning and obesity

Unlike cholesterol, certain fatty acids essential for growth, development and survival are not endogenously synthesized. These substances, vital components of cell membranes and precursors for prostaglandins, must be obtained from the diet. Humans require two classes of fatty acids, omega-3, found mostly in fish and shellfish, and omega-6, found primarily in vegetable oils (Connor & Connor, 1985). These are polyunsaturated fatty acids. The ratio of polyunsaturated to saturated fats in the diet is culturally and ecologically determined. A high ratio of polyunsaturated to saturated dietary fats is negatively associated with morbidity from arteriosclerosis and coronary heart disease, but may be positively associated with cancer (Assembly of Life Sciences, 1982). Although a non-genetic factor, populational differences in this ratio may affect expression of genetic propensities toward hyperlipidemia, vascular pathology, fat deposition and obesity.

Fat people have fat relatives whether they are genetically related or not (Garn, Cole & Bailey, 1979; Garn, Sullivan & Hawthorne, 1989). Furthermore, the probability that a parent or child will be obese is a function of the fatness of other family members. In addition, changes in fatness of spouses occur synchronously, indicating environmental effects on fatness (Garn *et al.*, 1979, 1989).

Heritability of obesity and skinfold (SF) fat pattern were examined with use of data from the NHLBI Twin Study, and showed a strong genetic influence on centrality of fat deposition (Selby *et al.*, 1989). Heritability for subscapular SF, but not triceps SF, was highly significant after adjusting subscapular SF for obesity with use of the body mass index (BMI). For overall obesity, as assessed with BMI or sum of skinfolds (subscapular + triceps skinfold) estimates of heritability were high, but only moderate after adjusting for other characteristics. Of all measures of fat and obesity examined, subscapular skinfold was most strongly associated with established risk factors for CVD (Selby *et al.*, 1989).

Analyses of familial resemblance of serial measures of BMI suggest a partial X-linked pattern of inheritance (Byard, Siervogel & Roche, 1989). Estimates using serial measurements are higher than those from cross-sectional family data, where weight and height were measured at different ages in parents and offspring. Such earlier data indicated increasing parent–child correlations with age; no such trend was observed when both generations were measured at the same chronological age (Byard *et al.*, 1989). Resting metabolic rate also shows familial aggregation and twin studies suggest that genetic variation may underlie some differences in energy storage and thereby influence obesity/adiposity (Lillioja *et al.*, 1986; McGarvey *et al.*, 1989).

In an ecological context of abundant calories and excessive ingestion, an increase in adiposity, whether measured as absolute weight, body mass index or skinfold thicknesses, is an expected outcome (McGarvey *et al.*, 1989). During recent decades, the focus of research into morbidity correlates of fatness and obesity has shifted from absolute amount of fat to sites of fat deposition or fat patterning (Vague, 1956; Mueller & Reid, 1979; Vague et al., 1985; Björntrop, 1988; Newell-Morris, Moceri & Fujimoto, 1989). Fat patterning, specifically a masculine or upper body distribution of fat, as opposed to total amount of fat may be the critical correlate of morbidity and mortality. Both twin and family studies have shown fat placement to be influenced by genetic factors. Furthermore, fat patterning may be under stricter genetic control than is total adiposity (Selby et al., 1989). This suggests that adipose tissue (energy stored as fat) may not have been detrimental to reproduction and/or survival during earlier phases of human evolution. Consequently, feedback mechanisms to inhibit energy deposition have not been necessary, since an excess was seldom available, and such mechanisms have not been coded into the human genome. Conversely, genes for placement of fat may have been necessary to ensure that stored energy was placed where it did not hamper vital functions. Today, fat deposition is observed in areas that are less metabolically and physiologically active, such as the buttocks, abdomen, dorsal aspect of the trunk, and subcutaneous spaces.

Obesity, observed in many populations today – Pima Indians (Knowler et al., 1981), Samoans (Pawson & Janes, 1982; Crews, 1988), Nauruans (Zimmet, 1979), the general population of the United States, Europe and European-derived populations (Keys, 1980; Lew & Garfinkel, 1979) – is unlikely to have characterized any pre-agricultural societies (Eaton et al., 1988). However, other than for NIDDM, a direct independent association of obesity, regardless of how measured, with mortality has not been firmly established (Andres, 1980; Keys, 1980; Crews, 1988, 1989). Still, in and of itself and regardless of any primary effect on metabolic, pulmonary or circulatory factors, massive obesity seems likely to place an increased mechanical burden on the cardiovascular system (see Borkan et al., 1986).

Glucose, hyperglycaemia and diabetes

Two major classes of diabetes mellitus are recognized. In both, excess plasma glucose is the sine qua non of diabetes mellitus. Insulin-dependent diabetes mellitus (IDDM) or type I, traditionally known as juvenile-onset diabetes, occurs at an early age, is accompanied by ketoacidosis and beta cell destruction, and responds to insulin therapy. Diabetes which occurs at later ages, is not accompanied by either symptom (at least in early stages), does not respond to exogenous insulin therapy, and appears in an obviously obese patient, is classified as non-insulin-dependent diabetes mellitus (NIDDM) or type II, traditionally known as adult-onset diabetes. Several rare forms of diabetes are also recognized and some of these are related to specific genetic defects (WHO, 1985; Foster, 1989).

Diabetes may be the most thoroughly examined of any chronic disease with respect to genetic etiology. IDDM is closely correlated with HLA haplotype, and beta cell destruction represents an autoimmune process related to a defect in self-recognition (Foster, 1989). That genetic factors were identified for IDDM and rare forms of diabetes suggests that NIDDM might also be secondary to a primary genetic defect. However, the search for a specific related genotype has not been fruitful. NIDDM is multifactorial in etiology, having both genetic and environmental causes (Leslie & Pyke, 1987). The genetic components are heterogeneous and probably involve several different loci (Xiang *et al.*, 1989). Genetic factors associated with glucose transport and metabolism and insulin secretion and resistance (for instance, insulin, glucagon, insulin receptor and apolipoprotein genes) have been examined in the hope of identifying primary genetic defects. HLA haplotypes have been examined because of their association with IDDM, but with generally negative results (Foster, 1989).

Apolipoprotein/lipid metabolism is often abnormal in diabetic patients (Laker, 1987). Primary variations in apolipoprotein and insulin-receptor (INS-R) genes are correlated with NIDDM in Chinese-Americans (Xiang *et al.*, 1989) and variation in apolipoprotein A-I (APO A-I) appears diabetogenic in Mexican-Americans (Hanis *et al.*, 1983). Synthesis of an abnormal INS-R also contributes to NIDDM in Pima Indians (Yamashita *et al.*, 1984). Interestingly, the human INS-R gene appears to be a mosaic, constructed of exons recruited from other sources, which has been poorly conserved in evolution (Seino *et al.*, 1989). This characteristic suggests the possibility of multiple variant alleles at the INS-R locus and the potential for rapid evolutionary change. The mosaic structure may also indicate previous gene amplification in response to selective pressures related to use of specific energy sources or adaptations to relatively low availability of energy sources during previous phases of human evolution.

INS-R, APO B and the APO A-I/C-III/A-IV gene cluster all contribute to development of NIDDM in Chinese-Americans (Xiang *et al.*, 1989). The INS-R locus may include a 'protective' phenotype, which acts regardless of other genes conferring susceptibility, and may identify groups of individuals at low risk of NIDDM. APO B apparently contributes to the development of NIDDM in persons of lean or normal weight, while the APO A-I/C-III/A-IV cluster may account for 8 % of additional risk, above baseline, in overweight individuals (Xiang *et al.*, 1989). Conversely, in black families segregating NIDDM, insulin and INS-R genes may not be major susceptibility loci, but still may contribute to a polygenic NIDDM phenotype (Cox, Epstein & Spielman, 1989).

A 'New World Syndrome' linking diabetes, upper body obesity, gall bladder disease and gallstones into a single metabolic unit has been postulated based on concordant increases that have occurred in these conditions in several Amerindian populations during the twentieth century (Weiss, Ferrell & Hanis, 1984). Population studies have shown that degree of Amerindian admixture is directly related to development of diseases in this 'metabolic unit' in Mexican-American and Amerindian samples (Chakraborty *et al.*, 1986). In some Amerindian and Mexican-American groups, overall serum cholesterol and HDL-cholesterol are lower and triglygerides are elevated compared with age- and sex-matched non-diabetic Euro-Americans (Weiss *et al.*, 1989). A similar pattern of cholesterol, triglycerides and HDL-cholesterol has been reported for Tarahumara Indians of Mexico (Connor *et al.*, 1978), Polynesians from the Samoan Islands (Pelletier & Hornick, 1986), and Yanomami Indians of North-western Brazil (Mancilha-Carvalho & Crews, 1990) in comparison to Euro-Americans. The indications are that these populations may metabolize fats, lipids and glucose differently from European-derived populations.

Evolutionary models have been proposed to explain current high prevalences of NIDDM in some populations. The best known is that proposed by J. V. Neel in 1962 (see also Neel, 1982). Briefly, evolution produced one or more 'thrifty' genotypes in some human populations exposed to fluctuating periods of scarcity and abundance in foodstuffs. In 1982, Neel restated this model and applied it only to NIDDM. Hypothetically, persons with 'thrifty' genes would have a selective advantage because of an enhanced ability to process food efficiently into energy and store it as fat. When the environment was low in caloric availability and there were periods of relative deprivation, 'thrifty' phenotypes would have had a selective advantage over individuals without these genes. They would rapidly develop hyperinsulinaemia with ingestion of large amounts of calories, and consequently would store more excess energy as adipose tissue. This phenotype would be exposed to selection during periods of food scarcity or starvation, and would be expected to outcompete other 'non-thrifty' phenotypes (Neel, 1962). In high calorie environments with little need for physical activity, these same phenotypes might be at a selective disadvantage. They would continuously experience hyperinsulinaemia, store more calories as adipose tissue and become obese, as is observed in many contemporary populations. NIDDM apparently is a disease of overabundance, exposed when evolutionarily determined mechanisms for conservation of scarce calories continually encounter a surfeit of calories. Although others have presented modified versions of Neel's original theory (Szathmary, 1985, 1989; Wendorf, 1989), none has yet contradicted the basic hypothesis.

Cardiovascular diseases

Cardiovascular diseases (CVD), particularly coronary heart disease (CHD), cerebral vascular disease (stroke), and myocardial ischemia, have been the object of intensive epidemiological and intervention research in the latter half of the twentieth century. Partly because CVD research designs have emphasized life-style and environmental rather than innate biological factors, relatively little is known of genetic factors predisposing to CVD in any population. Based on first degree relatives of probands with CHD, heritability of CHD has been estimated to be 20–80 % for men and 30–60 % for women (Robertson *et al.*, 1980). Study of 1000 adopted Danish children showed that risk of death prior to the age of 50 was double and risk of death from CVD was 4.5 times higher when a biological parent had died or died of CVD, respectively, prior to the age of 50 (Sørensen *et al.*, 1988). Familial aggregation of CHD and related risk factors is an established fact and genetic variability in lipid fractions and apolipoproteins are undoubtedly involved in the pathogenesis of atherosclerosis that presages CHD. For instance, of eight RFLPs identified in the APO A-II/C-III/A-IV cluster located on chromosome 11, six sites show relationships with phenotypic expression of coronary heart disease (Ferrell, 1989).

Cancer

A multifactorial basis, including susceptibility or oncogenes and environmental cofactors or triggers, for initiation and expression of neoplastic disease has long been hypothesized as the basis of most cancers (Peto, 1980; Doll & Peto, 1981). However, the environmental and genetic aetiology of a disease process often initiated as many as two or three decades prior to presentation of symptoms is difficult to determine. A number of factors are considered to be strongly cancer-promoting: high levels of dietary fat (or perhaps low fibre); alcohol and tobacco use;

aflotoxins and other mycotoxins; nitrates or nitrites in food; mutagens present in the environment and foods; and use of food additives.

Diet and environment are not simple factors to examine as are cigarette smoking, alcohol consumption, age or serum cholesterol (Hoff, Garruto & Durham, 1989). Several dietary factors appear to be protective against cancer, including dietary fibre, vitamins A, C, and E, selenium, beta-carotene-rich vegetables and cruciferous vegetables; whereas, salted, pickled and smoked foods, and fats appear to be cancer promoting (Assembly of Life Sciences, 1982). As to dietary fat, it may be that the source of ingested fat is more important than the absolute quantity. However, at least in persons subsisting on low-fat diets and contrary to CHD, polyunsaturated fats appear to be more tumourigenic than are the saturated fats (Assembly of Life Sciences, 1982).

Both tumour-promoting and tumour-suppressing genes may be wide-spread and segregating in human populations. The distribution of mye-loid leukaemia in men and unspecified leukaemia in women residing in Wyoming suggests the presence of recessive oncogenes, whereas the distribution of lung cancer suggests tumour-suppressing genes (Cleek, 1989). There is also evidence of tumour-suppressing genes for breast cancer. Cleek's (1989) work suggests the widespread existence of tumour-suppressing genes with *greater* effects on cancer distribution than tumour-inducing genes.

Early developmental factors and environment also appear influential in the aetiology of some cancers. A woman's age at the birth of her first child has shown a secular trend toward older ages in most United States and European populations since the nineteenth century, while age at menarche has declined (Micozzi, 1987). These trends are correlated with increased population breast cancer rates (Micozzi, 1987). This may reflect a misfit between present sociocultural circumstances and evolutio-narily determined aspects of reproduction in human females who are physiologically prepared for reproduction at an early age (13–14 years of age on average in westernized cosmopolitan societies: Beall, 1983). During past phases of human evolution, most women were probably pregnant and/or lactating soon after puberty and probably remained continuously pregnant or lactating throughout adulthood and into old age (Lancaster and King, 1985). Thus the hormonal context in which most of human evolution occurred probably was relatively high levels of circulat-ing progesterone and lactating hormones for most women, rather than the continued high oestrogen levels seen today in the majority of women who are non-pregnant and non-lactating over the largest proportion of their life-spans (Lancaster & King, 1985).

Another recent finding is that, during the development of some tumours, antigens appear that are expressed at various stages during early embryofoetal development, but not customarily in the postnatal period. These carcinoembryonic/oncofoetal antigens appear on the surface of embryofoetal and tumour cells, both of which are undergoing rapid growth (Hoff *et al.*, 1989). They probably are important as either growth factors or growth-associated cell receptors in tissue development in both embryofoetal and cancer genesis (Hoff *et al.*, 1989).

Discussion

Genetic epidemiology is a relatively young area of disease research that is closely related to human population biology. Their independent development at this time is particularly important because epidemiology, demography, clinical medicine and traditional biological anthropology have not established the aetiology of the major adult-onset chronic diseases that plague modern society. What these researches have revealed is that most adult-onset chronic conditions are multifactorial in origin and probably involve predispositions based in polygenetic rather than single-gene traits, interacting with a host of environmental and life-style factors which, until recently, have not been influential in shaping the human genome. Some such genetic factors may be inherited as variances in frailty or susceptibility to a spectrum of environmental factors (Vaupel, 1988). Others are likely to be traits related to single genes or gene complexes segregating as a single phenotypic condition with a constellation of characteristics. The aetiology of these conditions may best be examined using transdisciplinary research designs commonly used by genetic epidemiologists and human population biologists (Baker, 1982; Little & Haas, 1989).

Natural experimental designs in small anthropological populations that are relatively genetically, ecologically and culturally homogeneous provide natural laboratories for the study of health and disease, while avoiding some of the confounding factors that are inescapable in technologically complex cosmopolitan societies (Garruto *et al.*, 1989). This approach may have been best illustrated by studies of the aetiological basis of kuru among the Fore people of Papua New Guinea, that led to the discovery of a new category of virally transmitted diseases (Gajdusek, 1977). Other naturalistic experiments involve the study of migrants to determine environmental influences on diseases, while holding genetics constant through comparisons with their population of origin. These designs offer unique opportunities for determination of ecological and genetic basis of many major chronic diseases that plague modern societies (Garruto et al., 1989).

Results from migrants and population isolates may, however, not be directly generalizable to other populations. These populations are likely to possess unique gene pools or specific genes not duplicated outside the isolate (for example, Hutterites, Old-World Amish, Fore). Since multiple genotypes may lead to similar phenotypes, care must be taken when extrapolating to other populations. However, such studies do provide a basis for experiments in additional populations.

The epidemic of chronic degenerative diseases in present-day cosmopolitan populations has a multifactorial aetiology, rooted in joint interactions between susceptible genotypes, whether polygenic or single genes, and deleterious environmental factors. Recent demographic changes that have led to increased survival and life expectancy in most populations have exposed many of these characteristics, perhaps for the first time, in the nineteenth and twentieth centuries. It does not appear likely that any single disciplinary focus will unravel these complexities. Biological anthropology, and specifically human population biology with its transdisciplinary approach to the study of human disease processes, is in a position to contribute to an understanding of these multifactorial disease processes, NIDDM, CVD and cancer, which today account for over 70 % of all deaths in technologically complex cosmopolitan societies.

Chronic diseases may have a relationship to body form, morphology or other aspects of human variation traditionally examined by biological anthropologists, some of whom initially may have become interested in the prevalence and incidence of chronic diseases because of contributions of anthropometry to the study of those aspects. However, it is anthropologists' expertise about the relation of genetic to environmental factors in populations rather than in single individuals that may provide the most significant area for future application of biological anthropology to work on these human problems.

Acknowledgements
This manuscript was prepared while DEC was partly supported by Grant RO1 AG08395 from the National Institute on Aging. The assistance of Evelyn Howard Sala with typing is gratefully acknowledged. Our thanks are due also to G. Lasker, who commented usefully on an earlier version of this manuscript.

References
Adams, J. & Smouse, P. E. (1985). Genetic consequences of demographic changes in human populations. In Diseases of Complex Etiology in Small

Populations, ed. R. Chakraborty & E. J. E. Szathmary, pp. 147–78. New York: Alan R. Liss.

AHA (American Heart Association) (1988). The Joint National Committee on Detection, Evaluation, and Treatment of High Blood Pressure, The 1988 report. *Archives of Internal Medicine*, **148**, 1020–38.

Andres, R. (1980). Effect of obesity on total mortality. *International Journal of Obesity*, **4**, 381–6.

Assembly of Life Sciences, Committee on Diet, Nutrition, and Cancer, NRC (1982). *Diet, Nutrition and Cancer*. Washington, DC: National Academy Press.

Baker, P. T. (1982). Human population biology: a viable transdisciplinary science. *Human Biology*, **54**, 203–20.

(1984). The adaptive limits of human populations. *Man* (NS) **19**, 1–14.

Baker, P. T. & Baker, T. S. (1977). Biological adaptations to urbanization and industrialization: some research strategy considerations. In *Colloquia in Anthropology, volume I*, ed. R. K. Wetherington, pp. 107–18. Fort Burgwin Research Center.

Beall, C. M. (1983). Ages at menopause and menarche in a high altitude Himalayan population. *Annals of Human Biology*, **10**, 365–70.

Björntrop, P. (1988). Possible mechanisms relating fat distribution and metabolism. In *Fat Distribution During Growth and Later Health Outcomes*, ed. C. Bouchard & F. Johnson, pp. 175–91. New York: Alan R. Liss.

Boerwinkle, E., Turner, S. T., Weinshilboun, R., Johnson, M., Richelson, E. & Sing, C. F. (1986). Analysis of the distribution of erythrocyte sodium-lithium countertransport in a sample representative of the general population. *Genetic Epidemiology*, **3**, 365–8.

Borkan, G., Sparrow, D., Wisniewski, C. & Vokonas, P. S. (1986). Body weight and coronary disease risk: patterns of risk factor change associated with long-term weight gain. *American Journal of Epidemiology*, **124**(3), 410–19.

Brown, M. S. & Goldstein, J. L. (1989). Familial hypercholesterolemia. In *The Metabolic Basis of Inherited Disease* (6th edn), ed. C. R. Scriver, A. L. Beaudet, W. S. Sly & D. Valle, pp. 1215–50. New York: McGraw-Hill.

Byard, P. J., Siervogel, R. M. & Roche, A. F. (1989). X-linked pattern of inheritance for serial measures of weight/stature^2. *American Journal of Human Biology*, **1**, 443–9.

Chakraborty, R., Ferrell, R., Stern, M. P., Haffner, S. M., Hazuda, H. P. & Rosenthal, M. (1986). Relationship of prevalence of non-insulin-dependent diabetes mellitus to Amerindian admixture in the Mexican Americans of San Antonio, Texas. *Genetic Epidemiology*, **3**, 435–54.

Christlieb, A. R., Krolewski, A. S., Warran, J. H. & Soeldner, J. S. (1985). Is insulin the link between hypertension and obesity. *Hypertension*, **7** (Supplement II), 54.

Cleek, R. K. (1989). Surnames and cancer genes. *Human Biology*, **61**(2), 195–211.

Connor, W. E., Cerqueria, M. T., Rodney, M. S., Connor, R. W., Walla, R. B., Malinow, M. R. & Casdorph, H. R. (1978). The plasma lipids, lipoproteins, and diet of Tarahumara Indians of Mexico. *American Journal of Clinical Nutrition*, **31**, 1131–42.

Connor, W. E. & Connor, S. L. (1985). The dietary prevention and treatment of coronary heart disease. In *Coronary Heart Disease: prevention, complications, and treatment*, ed. W. E. Connor & J. D. Bristow, pp. 43–64. Philadelphia: J. B. Lippincott.

Cox, N. J., Epstein, P. A. & Spielman, R. S. (1989). Linkage studies on NIDDM and the insulin and insulin-receptor genes. *Diabetes*, 38, 653–8.

Crews, D. E. (1988). Body weight, blood pressure and the risk of total and cardiovascular mortality in an obese population. *Human Biology*, 60, 417–33.

— (1989). Multivariate prediction of total and cardiovascular mortality in an obese Polynesian population. *American Journal of Public Health*, 79(8), 982–6.

Davignon, J., Gregg, R. E. & Sing, C. F. (1988). Apolipoprotein E polymorphism and arteriosclerosis. *Arteriosclerosis*, 8, 1–21.

Doll, R. & Peto, R. (1981). The causes of cancer: quantitative estimates of avoidable risks of cancer in the United States. *Journal of the National Cancer Institute*, 66, 1193–306.

Eaton, S. B. & Konner, M. J. (1985). Paleolithic nutrition: a consideration of its nature and current implications. *New England Journal of Medicine*, 312, 283–9.

Eaton, S. B., Konner, M. J. & Shostak, M. (1988). Stoneagers in the fastlane: chronic diseases in evolutionary perspective. *American Journal of Medicine*, 84, 739–49.

Falconer, D. S. (1960). *Introduction To Quantitative Genetics*. Edinburgh: Oliver & Boyd.

Ferrell, R. E. (1989). Application of molecular techniques to the study of human physiological variation. *American Journal of Human Biology*, 1, 545–53.

Foster, D. W. (1989). Diabetes mellitus. In *The Metabolic Basis of Inherited Disease* (6th edn), ed. C. R. Scriver, A. L. Beaudet, W. S. Sly & D. Valle, pp. 375–97. New York: McGraw-Hill.

Gajdusek, D. C. (1977). Unconventional viruses and the origin and disappearance of kuru. In *Les Prix Nobel en 1976* (Nobel Foundation), pp. 167–216. Stockholm: P. A. Norstedt & Soner.

Garn, S. M., Cole, P. E. & Bailey, S. M. (1979). Living together as a factor in family-line resemblances. *Human Biology*, 51, 565–87.

Garn, S. M., Sullivan, M. T. & Hawthorne, V. A. (1989). Educational level, fatness, and fatness differences between husbands and wives. *American Journal of Clinical Nutrition*, 50, 740–5.

Garruto, R. M., Way, A. B., Zansky, S. & Hoff, C. (1989). Natural experimental models in human biology, epidemiology, and clinical medicine. In *Human Population Biology: a transdisciplinary science*, ed. M. A. Little & J. D. Haas, pp. 82–109. New York: Oxford University Press.

Glueck, C. J., Laskarzewski, P. M., Rao, D. C. & Morrison, J. A. (1985). Familial aggregation of coronary risk factors. In *Coronary Heart Disease: prevention, complications, and treatment*, ed. W. E. Connor & J. D. Bristow, pp. 173–92. Philadelphia: J. B. Lippincott.

Green, R., Owen, A. R. G., Namboodiri, K., Hewitt, D., Williams, L. R. & Elston, R. C. (1984). The Collaborative Lipid Research Program

Family Study: detection of major gene influencing lipid levels by examination of heterogeneity of familial variances. *Genetic Epidemiology*, 1, 123–41.

Hanis, C. L., Ferrell, R. E., Barton, S. A., Aguilar, L., Garza-Ibarra, A., Tulloch, B. R., Garcia, C. A. & Schull, W. J. (1983). Diabetes among Mexican-Americans in Starr Country, Texas. *American Journal of Epidemiology*, 118, 659–72.

Harrison, G. A. (1973). The effects of modern living. *Journal of Biosocial Science*, 5, 217–28.

Hinkle, L. E. (1987). Stress and disease: the concept after 50 years. *Social Science and Medicine*, 25, 561–6.

Hoff, C., Garruto, R. M. & Durham, N. M. (1989). Human adaptability and medical genetics. In *Human Population Biology: a transdisciplinary science*, ed. M. A. Little & J. D. Haas, pp. 69–81. New York: Oxford University Press.

Illingworth, D. R. & Connor, W. E. (1985). Hyperlipidemia and coronary heart disease. In *Coronary Heart Disease: prevention, complications, and treatment*, ed. W. E. Connor & J. D. Bristow, pp. 21–42. Philadelphia: J. B. Lippincott.

INTERSALT Cooperative Research Group (1988). INTERSALT: an international study of electrolyte excretion and blood pressure. Results for 24-hour urinary sodium and potassium excretion. *British Medical Journal*, 297, 319–30.

James, G. D. & Baker, P. T. (1990). Human population biology and hypertension: evolutionary and ecological aspects of blood pressure. In *Hypertension: pathophysiology, diagnosis, and management*, ed. J. H. Laragh & B. M. Brenner, pp. 137–45. New York: Raven Press.

James, G. D., Crews, D. E. & Pearson, J. D. (1989). Catecholamines and stress. In *Human Population Biology: a transdisciplinary science*, ed. M. A. Little & J. D. Haas, pp. 280–95. New York: Oxford University Press.

Keys, A. (1980). Overweight, obesity, coronary heart disease and mortality. *Nutrition Reviews*, 38, 297–307.

Knowler, W. C., Pettit, D. J., Savage, P. J. & Bennett, P. H. (1981). Diabetes incidence in Pima Indians: Contributions of obesity and parental diabetes. *American Journal of Epidemiology*, 113, 144–56.

Laker, M. F. (1987). Plasma lipids and apolipoproteins in diabetes mellitus. In *The Diabetes Annual/3*, ed. K. G. M. M. Alberti & L. P. Krall, pp. 459–78. Amsterdam: Elsevier.

Lancaster, J. B. & King, B. J. (1985). An evolutionary perspective on menopause. In *In Her Prime: a new view of middle-aged women*, ed. J. K. Brown & V. Kerns. South Hadley: Bergin & Garvey.

Laragh, J. H. & Brenner, B. M. (1990). *Hypertension: pathophysiology, diagnosis, and management*. New York: Raven Press.

Leslie, R. D. G. & Pyke, D. A. (1987). Genetics of diabetes. In *The Diabetes Annual/3*, ed. K. G. M. M. Alberti & L. P. Krall, pp. 39–54. Amsterdam: Elsevier.

Levi, L. & Anderson, L. (1975). *Psychosocial Stress*. New York: Spectrum Publications.

Lew, E. A. & Garfinkel, L. (1979). Variations in mortality by weight among 750 000 men and women. *Journal of Chronic Disease*, **32**, 563–76.

Lillioja, S., Ravussin, E., Abboyy, W., Zawadski, J. K., Young, A., Knowler, W. C., Jacobowitz, R. & Moll, P. P. (1986). Familial dependency of resting metabolic rate. *New England Journal of Medicine*, **315**, 96–100.

Little, M. A. & Haas, J. D. (eds) (1989). *Human Population Biology: a transdisciplinary science*. New York: Oxford University Press.

Lucas, C. P., Estigarribia, J. A., Darga, L. L. & Reaven, G. M. (1985). Insulin and blood pressure in obesity. *Hypertension*, **7**, 702–6.

McGarvey, S. T., Bindon, J. R., Crews, D. E. & Schendel, D. E. (1989). Modernization and adiposity: causes and consequences. In *Human Population Biology: a transdisciplinary science*, ed. M. A. Little & J. D. Haas, pp. 263–79. New York: Oxford University Press.

Mancilha-Carvalho, J. J. & Crews, D. E. (1990). Lipid profiles of Yanomamo Indians in Brazil. *Preventive Medicine*, **19**(1) (in press).

Micozzi, M. (1987). Cross-cultural correlations of childhood growth and adult breast cancer. *American Journal of Physical Anthropology*, **73**, 525–37.

Morton, N. E., Gulbrandsen, C. L., Rao, D. C., Rhoads, C. G. & Kagan, A. (1980). Determinants of blood pressure in Japanese-American families. *Human Genetics*, **53**, 261–6.

Mueller, W. H. & Reid, D. M. (1979). A multivariate analysis of fatness and relative fat patterning. *American Journal of Physical Anthropology*, **50**, 199–208.

Neel, J. V. (1962). Diabetes mellitus: a 'thrifty genotype' rendered detrimental by progress? *American Journal of Human Genetics*, **14**, 353–62.

(1982). The thrifty genotype revisited. In *The Genetics of Diabetes Mellitus*, ed. J. Kobberling & R. B. Tattersall, pp. 283–93. New York: Academic Press.

Newell-Morris, L., Moceri, V. & Fujimoto, W. (1989). Gynoid and android fat patterning in Japanese-American men: body build and glucose metabolism. *American Journal of Human Biology*, **1**, 73–86.

Pawson, I. G. & Janes, C. (1982). Massive obesity in a migrant Samoan population. *American Journal of Public Health*, **71**, 508–13.

Pelletier, D. & Hornick, C. (1986). Blood lipid studies. In *The Changing Samoans: behavior and health in transition*, ed. P. T. Baker, J. M. Hanna & T. S. Baker, pp. 327–49. Oxford: Oxford University Press.

Peto, J. (1980). Genetic predisposition to cancer. In *Cancer Incidence in Defined Populations*, ed. J. Cairus, J. Lyon & M. Skolnick, pp. 203–13. Cold Spring Harbor, New York: Cold Spring Harbor Laboratory.

Pickering, G. W. (1961). *The Nature of Essential Hypertension*, New York: Grune & Stratton.

(1967). The inheritance of arterial pressure. In *The Epidemiology of Hypertension*, ed. J. Stamler, R. Stamler & T. N. Pullman, pp. 18–27. New York: Grune & Stratton.

Pickering, T. G., Harshfield, G. A., Devereux, R. B. & Laragh, J. H. (1985). What is the role of ambulatory blood pressure monitoring in the management of hypertensive patients? *Hypertension*, **7**, 171–7.

Robertson, F. W., Cumming, A. M., Douglas, A. S., Smith, E. B. & Kemmure, A. C. (1980). Coronary heart disease in North-east Scotland: a study of genetic and environmental variation in serum lipoproteins and other variables. *Scottish Medical Journal*, **25**, 212–21.

Seino, S., Seino, M., Nishi, S. & Bell, G. I. (1989). Structure of the human insulin receptor gene and characterization of its promoter. *Proceedings of the National Academy of Sciences*, **86**, 114–18.

Selby, J. V., Newman, B., Queensbury, C. P. Jr, Fabsitz, R. R., King, M.-C. & Meaney, F. J. (1989). Evidence of genetic influence of central body fat in middle-aged twins. *Human Biology*, **61**, 179–93.

Sing, C. F., Boerwinkle, E. & Moll, P. P. (1985). Strategies for elucidating the phenotypic and genetic heterogeneity of a chronic disease with a complex etiology. In *Diseases of Complex Etiology in Small Populations*, ed. R. Chakraborty & E. J. E. Szathmary, pp. 39–66. New York: Alan R. Liss.

Sørensen, T. I. A., Nielsen, G. G., Andersen, P. K. & Twasdale, T. W. (1988). Genetic and environmental influences on premature death in adult adoptees. *New England Journal of Medicine*, **318**, 727–32.

Stini, W. A. (1971). Evolutionary implications of changing nutritional patterns in human populations. *American Anthropologist*, **73**, 1019–30.

Szathmary, E. J. E. (1985). The search for genetic factors controlling glucose levels in Dogrib Indians. In *Diseases of Complex Etiology in Small Populations*, ed. R. Chakraborty & E. J. E. Szathmary, pp. 199–226. New York: Alan R. Liss.

(1989). The impact of low carbohydrate consumption on glucose tolerance, insulin concentration and insulin response to glucose challenge in Dogrib Indians. *Medical Anthropology*, **11**, 329–50.

Trowell, H. & Burkitt, D. P. (eds) (1981). *Western Diseases: their emergence and prevention*. Cambridge, Mass.: Harvard University Press.

Vague, J. (1956). The degree of masculine differentation of obesities: a factor determining predisposition to diabetes, atherosclerosis, gout, and uric acid calculous disease. *American Journal of Clinical Nutrition*, **4**, 20–34.

Vague, J., Björntrop, P., Guy-Grand, B., Rebuffe-Scrive, M. & Vague, P. (1985). *Metabolic Complications of Human Obesities*. Amsterdam: Excerpta Medica.

Voupel, J. W. (1988). Inherited frailty and longevity. *Demography*, **25**, 277–87.

Ward, R. H. (1985). Isolates in transition: a research paradigm for genetic epidemiology. In *Diseases of Complex Etiology in Small Populations*, ed. R. Chakraborty & E. J. E. Szathmary, pp. 147–78. New York: Alan R. Liss.

(1990). Familial aggregation and genetic epidemiology of blood pressure. In *Hypertension: pathophysiology, diagnosis, and management*, ed. J. H. Laragh & B. M. Brenner, pp. 81–100. New York: Raven Press.

Weiss, K. M., Ferrell, R. F. & Hanis, C. L. (1984). A New World Syndrome of metabolic diseases with a genetic and evolutionary basis. *Yearbook of Physical Anthropology*, **27**, 153–78.

Weiss, K. M., Ulbrecht, J. S., Cavanagh, P. R. & Buchanan, A. V. (1989). Diabetes mellitus in American Indians: characteristics, origins and preventive health care implications. *Medical Anthropology*, **11**, 283–304.

Wendorf, M. (1989). Diabetes, the ice free corridor, and the Paleoindian settlement of North America. *American Journal of Physical Anthropology*, **79**, 503–20.

WHO Study Group (1985). *Diabetes Mellitus: Technical Report Series* 727. Geneva: WHO.

Williams, B. J. (1973). *Evolution and Human Origins*. New York: Harper & Row.

Williams, G. C. (1957). Pleiotropy, natural selection, and the evolution of senescence. *Evolution*, 11, 398–411.

Wilson, T. W. (1986). Africa, Afro-Americans, and hypertension: an hypothesis. *Social Science History*, 10, 489–98.

Yamashita, T., Mackay, W., Rushforth, N., Bennett, P. & Houser, H. (1984). Pedigree analysis of non-insulin dependent diabetes mellitus (NIDDM) in the Pima Indians. *American Journal of Human Genetics*, 36, 18–35.

Yunis, J. J. & Hoffman, W. R. (1989). Nuclear enzymes, fragile sites, and cancer. *Journal of Gerontology: Biological Sciences*, 44, 37–44.

Xiang, K.-S., Cox, N. J., Sanz, N., Huang, P., Karam, J. H. & Bell, G. I. (1989). Insulin-receptor and apolipoprotein genes contribute to development of NIDDM in Chinese Americans. *Diabetes*, 38, 17–23.

Zimmet, P. (1979). Epidemiology of diabetes and its macrovascular manifestations in Pacific populations: the medical effects of social progress. *Diabetes Care*, 2, 144–53.

8   *The biology of human aging*

WILLIAM A. STINI

Introduction

Biological anthropologists have long been interested in human variation
along with the processes of growth, development and maturation that
produced the highly individualistic phenotypes that make up human
populations. Other disciplines have also concerned themselves with the
phenomenon of human variation, sometimes from the perspective of its
relation to human performance as is the case with sports medicine or
exercise physiology. Still others, such as paediatrics and human nutrition,
focus on the factors that influence growth and maintenance. Biological
anthropologists, however, approach the study of human variability from
an evolutionary perspective. The questions that are generated by the
evolutionary approach will often yield less precise answers but have the
redeeming virtue of opening up areas of investigation that might otherwise be ignored. Because of this ability to see problems from an
evolutionary perspective, the biological anthropologist has much to
contribute to the study of human aging.

For instance, world-wide demographic trends are reshaping the demographic profile in unprecedented ways. A large proportion of the human
population is surviving beyond the age of reproduction. Most notable is
the extension of life expectancy of women who, in industrialized nations
of North America, Europe and Japan, may expect to outlive men by an
average of 6 or 7 years. Yet it is the female of our species who experiences
a clear-cut end to reproductive life with the completion of menopause.
What is the evolutionary significance of this seemingly paradoxical
observation? It appears that superior life expectancies are a characteristic
of the females of many species (Rockstein, Chesky & Sussman, 1977).
The reasons for this superiority are yet to be determined. However,
where life expectancies in other species have been ascertained with
acceptable levels of accuracy, it has usually been through observation of
domesticated individuals. Seldom do free-ranging animals survive long
enough to reach the limits of senescence (Snyder & Moore, 1968;
Western, 1979; Jones, 1982; Western & Ssemakula, 1982; Cutler, 1984).

Humans are quite literally a self-domesticated species. The dramatic extension of human life expectancies that has occurred over the last 100 years is largely attributable to the success of the technological innovations that buffer the individual from environmental stress. At the same time that life expectancies have been increasing, human populations have also increased their control over reproduction. The net result has been a restructuring of the demographic profile so that it little resembles the broad-based pyramidal form seen in non-industrialized areas and, presumably, the one characteristic of our species for most of its history.

The emergent demographic trend for the entire human population is, as far as is presently known, unprecedented in the history of life on this planet. Therefore, our species is the first to experience the process of senescence to death on a large scale. Since idiosyncratic differences between individuals are greatest among the old, students of human variation will be increasingly engaged in the observation of this unprecedented natural experiment. The study of human aging intersects all of the biomedical sciences and has much to offer to each. Since it is concerned with the description and, ultimately, the explanation of a new chapter in the history of life on earth, it has much theoretical as well as practical potential. Biological anthropologists may rightfully claim a central role in the development of research of both basic and applied significance.

Why do aging and senescence occur?

Are senescence and death a necessary concomitant of life? This is a question that has engaged biologists for centuries. Over 100 years ago, Weismann (1889) argued that death is not a necessary attribute of life but was acquired secondarily as an adaptation. He argued that a limited lifespan was adaptive: 'not because it is contrary to its nature to be unlimited but because the unlimited existence of individuals would be a luxury without any corresponding advantage'. However, senescence and death are not mandatory for all living species. For instance, bacteria and haploid species of *Protozoa* living today enjoy apparent immortality. They divide and redivide with no detectable decline in the ability to do so. Nor do they exhibit decreased metabolic activity or structural changes that might indicate senescence. According to Smith-Sonneborn (1987), the loss of ability to multiply forever first occurred in eukaryotic protozoans. It is believed that the first eukaryotes appeared sometime between 1.2 and 1.4 billion years ago, some 2 billion years after the first prokaryotes. Consequently, aging is thought to be an evolved trait dating back as much as a billion years. Therefore, it appears that senescence evolved in lines where the more primitive ancestors did not age (Hirsch, 1987).

The evolution of senescence followed the emergence of sexual reproduction which made possible recombination of genetic traits through exchange of multiple chromosomes by individuals possessing diploid sets of chromosomes. Gamete formation through the process of meiosis was an essential element of the evolution of the eukaryote mode of reproduction. An important event preceding attainment of this capacity was the appearance of a specialized nucleus within a given cell. This may have occurred as seen in present-day ciliates, which possess a germ-line micronucleus and a somatic line micronucleus. Alternatively, sexual reproduction may have arisen as a result of cell specialization producing reproductive cells on the one hand and somatic cells on the other, as seen in present-day colonial flagellates such as *Volvox*. Whichever pathway was followed, the emergence of a dichotomous system wherein an immortal germ line perpetuated itself through the exploitation of a disposable somatic line set the stage for the appearance of senescence and death. Kirkwood & Cremer (1982) have introduced the concept of the 'disposable soma' to describe this process. Kirkwood argues (1977, 1981, 1985, 1987, 1989) that specialization of cells is incompatible with indefinite survival. One reason for this is that intrinsic metabolic accidents will inevitably lead to cell mortality, as is seen in post-mitotic cells of higher organisms. Proliferating cells may have evolved finite replicative lifespans as protection against such disorders as cancer. The ability to separate the germ line from all lines that could ultimately threaten its survival favoured the evolution of traits that would programme senescence and death for the disposable somatic line.

Once this separation occurred, the advantages of outbreeding in facilitating the accumulation of genetic variability influenced the direction of evolution. Outbreeding organisms can achieve genetic variability comparable to that of inbreeding ones with a much lower mutation rate, hence lower risk of a lethal alteration of metabolic processes (Nanney, 1974). Ideally, outbreeding species favour mating patterns involving mate selection from strangers, a system which facilitates the accumulation of genetic variety. To promote outbreeding, such species have different mating types, enhancing the chances of a stranger being of a different mating type and therefore suitable as a sexual partner. Another characteristic associated with emphasis on outbreeding is lengthening of the period of immaturity. A long period of immaturity increases the likelihood of spatial separation of closely related individuals before mating occurs, further reducing the chance of inbreeding (Smith-

Sonneborn, 1987). The enhancement of variability effected through outbreeding is reinforced by recombinations during the meiotic cell divisions of gametogenesis. Moreover, meiosis can be a time of highly asymmetric partitioning of cellular defects, resulting in the elimination of non-viable genotypes by 'dumping' them in polar bodies which ultimately die (Sheldrake, 1974). Thus, the mechanisms of sexual reproduction characteristic of higher organisms simultaneously ensure continuing reassortment of traits, while providing an avenue for elimination of deleterious genes during the comparatively inexpensive pre-zygotic stage of the process (Bernstein, 1977).

While the evolution of sexual reproduction made senescence and death necessary, the evolutionary course set by increasing the period of immaturity potentiated a trend toward extended life-span in a number of phyletic lines. The interplay between the 'immortal' germ line and the 'disposable' somatic line is modulated to maximize individual survival in a manner that ensures production of a number of progeny appropriate for the continuation of the species. With extension of the period of immaturity came an increase in parental investment in longer-lived species (Strehler, 1977), The corollary of increased investment in individual progeny was a reduction in the number of progeny per parent. To quote Kirkwood (1987):

> One way to view an organism is as an entity which takes in energy from its environment in the form of nutrients and other resources and eventually produces as output its progeny. The organism must allocate a fraction of the energy taken in to each of the various somatic activities such as growth, defense, repair, and maintenance activities as well as reproduction. The balance between these activities is a central part of optimizing the life history. The greater the fraction of energy invested in one particular activity, the less will remain for the others.

Optimizing strategies and life-span

Examination of the energy allocations made by various organisms reveals a variety of optimizing strategies. However, it is possible to view the range of prioritizing strategies as divided into two broad but overlapping categories. These categories may be defined, following Pianka (1974) as the r strategy and the K strategy. The r strategist maximizes the number of progeny produced and minimizes the amount of energy allocated for the various somatic activities. The K strategist limits the energy allocated to reproduction by calibrating the number of progeny produced to the carrying capacity of the environment. Although there are probably no

Table 8.1. r *and* K *selection compared*

Correlate	r selection	K selection
Climate	Unpredictable	Predictable
Mortality	Catastrophic, density-independent	Density-dependent
Population size	Variable, below carrying capacity	Constant, near carrying capacity
Competition	Variable, not stringent	Usually intense
High selective value	1 High maximal rate of increase 2 Early reproduction 3 Rapid development 4 Small body size	1 Greater competitive ability 2 Delayed reproduction 3 Slower development 4 Larger body size
Length of life	Short	Longer
Chief advantage	Productivity	Efficiency

Adapted from Pianka (1974).

pure r strategists or K strategists, the attributes that would be associated with each strategy can be usefully compared for the purpose of assessing the predominant strategy of a given species. Table 8.1, following Pianka's designations, provides a listing of such attributes (Pianka, 1974).

It is clear from the list of attributes seen in Table 8.1 that the constellation of traits characterizing the K strategist has the unifying theme of preservation of the individual through enhanced investment in each of the progeny through an extended period of growth, maturation, training and, in our own species, enculturation. How can this emphasis on the preservation of the individual be reconciled with the indisputable existence of a limited life-span for each species?

It has been shown through laboratory experiments with *Drosophila* that a species-specific life span can be significantly increased through selection (Arking, 1987). When such increases occur, it is through extension of the pre-senescent segment of the life cycle, not of the length of time between the onset of senescence and death. It therefore appears that the factors that trigger the onset of senescence are the best candidates for the role of limiters of life-span. It is likely that relatively few genes are responsible (Cutler, 1975). According to Arking, there may indeed be a small number of aging mechanisms shared by all living forms. These would be analogous to the 'homeobox' that governs the developmental process. In *Drosophila*, the genes for the expression of the long-lived phenotype have been traced to the second chromosome, although they appear to be modulated by genes located on the third chromosome. If one accepts the premise that a life-limiting homeobox is widely shared among vertebrates and invertebrates, the antiquity of the genetic predisposition to senescence and death far exceeds that of traits which have combined to define the K strategist. Thus, the K strategy may achieve a certain degree of success in delaying the onset of senescence and death, but cannot overcome the overriding necessity for species to evolve through natural selection.

Reinforcement of the tendency to retain senescence and death even among K strategists comes from the occurrence of positive selection value for traits that yield beneficial effects early in life but, as a pleiotropic consequence, reduce fitness later in life. An attribute that conferred a survival advantage in the earlier part of the life-span when survivorship in the natural environment was high would be favoured by selection even if that same attribute became disadvantageous at a later stage in life when survivorship was low (Williams, 1957). Even in the absence of an intrinsic aging process, the force of natural selection is progressively attenuated with increasing age, since the fraction of survivors steadily decreases (Haldane, 1941; Medawar, 1952). Thus, selection for traits favourable to extension of the life-span under natural conditions would not be sufficiently strong to offset the factors favouring senescence and death.

It is of considerable interest that the changes associated with senescence are remarkably similar in organisms as distantly related as mammals and nematodes (Russell & Seppa, 1987). In both phyletic lines, senescence is marked by declining motor performance, accumulation of the so-called 'aging pigment' lipofuscin and a characteristic population survival curve. Decreasing fecundity and length of time of sustained exercise are traits shared by a wide range of phyla. In houseflies, for instance, the length of time in flight decreases proportionately to increase in age beyond the fourth day of life (Sohal & Runnels, 1986).

Cell death and aging

Although senescence and death of organisms share many characteristics across a wide spectrum of phyla, it is also true that cell death is not restricted to the senescent period of life. Many tissues maintain themselves through continual death and replacement of their constituent cells. Some epithelial cells, such as those in the human epidermis, may serve their purpose after dying, the cornified layer of human epidermis being made up of cells that have migrated to the surface and ceased all metabolic activity over an approximately 28-day period. Orgel (1973) argued that errors in DNA replication ultimately accumulated to a point where cell function failed. Human red blood cells have a life-span of approximately 120 days, most of which follows the loss of the nucleus. Even the process of embryogenesis involves cell death. In the chick embryo, an area in the vicinity of the juncture of the wing and body is programmed to die at a specific time in development. The cells of this region (called the 'posterior necrotic zone') die as a result of the breakdown of lysosomal membranes, an event which occurs whether the cells are *in vivo* or *in vitro* (Cunningham & Brookbank, 1988). In human embryogenesis, neurons of the spinal cord are lost during early development (Hamburger, 1958), and cells die and are resorbed as part of the process producing separated fingers in the embryonic hand (Saunders, 1966). Similarly, programmed cell death has also been described in invertebrates (Grigliatti, 1987).

Limitations on the ability to synthesize and repair DNA appear to limit the life-span of many cells in eukaryotic organisms. Hayflick's observation of a finite number of divisions in cultured human fibroblasts has been confirmed by a number of other researchers, although cancer cells do not seem to be subject to such limitations (Hayflick, 1985). Terminally non-dividing human fibroblasts have been shown to inhibit DNA synthesis in normal and some immortal cells that would otherwise be capable of DNA synthesis and continued cell division (Norwood *et al.*, 1974; Stein & Yanishevsky, 1979). It has been shown that such inhibitory activity can be prevented by treating the inhibiting cells with trypsin. From this observation, Periera-Smith, Fisher & Smith (1985) have concluded that senescent cells have an inhibiting protein present on the surface of their plasma membranes. Smith, Spiering and Periera-Smith (1987) also showed that when normal human fibroblasts were hybridized with immortal cells, the hybrids had a limited life of as much as 70 cell divisions, indicating that limited life is dominant and immortality recessive. Drescher-Lincoln & Smith (1983) demonstrated that cytoplasts from terminally non-dividing human fibroblasts could inhibit DNA synthesis by other cells as effectively as whole cells. Similar observations had also been made by Burmer *et al.* (1983). Drescher-Lincoln & Smith (1984) went on to treat senescent cytoplasts with cycloheximide or puromycin prior to fusion and effectively eliminated their inhibitory capability. Both of these treatments inhibit protein synthesis, adding evidence that inhibition is effected by a protein produced by senescent cells. However, when senescent human cells are infected with a cancer-inducing virus (simian virus 40), DNA synthesis is initiated, indicating that under appropriate circumstances inhibition is reversible (Ide *et al.*, 1983).

While inability to synthesize DNA and divide could be the chief determinant of senescence and death in dividing cells, such as fibroblasts, loss of capability to repair DNA could be more important in post-mitotic cells, such as neurons. Many, if not all, cells are subjected to damage by ionizing radiation such as ultraviolet light (Francis, Lee & Regan, 1981). The effect of such damage is similar to that of somatic mutations, resulting in either loss of gene products or cell death (Hirsch, 1978). Bergmann *et al.* (1981) showed that ultraviolet radiation induced DNA damage in primate lymphocytes and that the ability to repair such damage was correlated with the cells' life-span. In general, long-lived species, such as elephants and humans, are characterized by superior DNA-repair capacities, while short-lived ones, such as mice, shrews, rats, and hamsters, have poor DNA repair. Species with intermediate life-spans, such as the domestic cow, also exhibit intermediate DNA repair capacity (Hart & Turturro, 1983).

The consequences of failure to repair DNA damage vary with respect to the affected tissues and functional impairments. An area of major significance is that of immune competence. One important aspect of human aging is the eventual decline of the effectiveness of the immune system. It has not yet been demonstrated that failure of the immune system is the result of a decline in DNA repair. However, the importance of DNA splicing and recombination to the immune response makes it an area of great vulnerability to factors depressing DNA synthesis. Johnson (1988) argues that the major histocompatibility complex (HLA in humans) is a supergene system that regulates both the immune system and the aging process. The extent to which declining DNA synthesis and repair in this system are implicated in immune failure in old age remains an intriguing question. In addition, the relationship of accumulated products of metabolism such as free radicals to impaired DNA synthesis and repair deserves continued attention (Balin, 1982).

Comparative life-spans of living species

Is there a clue to the determinants of longevity to be found in the comparative life-spans of contemporary species? If there is, it remains to be recognized. One vexing problem in the comparative approach is that there is such a wide range of variability in life-spans within phyla. There appear to be long- and short-lived species throughout both the plant and animal kingdoms, and it has not been possible to identify which properties differentiate them in any general sense, although some attempts to do so have raised interesting questions which will be addressed in a

Table 8.2. *Maximal life-spans*

Organism	Approximate maximum age (years)
Bristlecone pine	5000
Sequoia	4000–5000
Italian cypress	2000
Tortoise	170
Rockfish	140
Humans	120
Sea anemones	90
Blue whale	80
Indian elephant	70
Owl	65
Horse	60
Alligator	56
Sponge	50
Termite	40
Gorilla	39
Domestic dog	34
Canada goose	32
Rhesus monkey	29
Domestic cat	28
Leech	27
Grey squirrel	15
Tarantula	15
Rabbit	13
Honey bee queen	6
Rat	5
Shrew	2

subsequent section. Table 8.2 (modified from Cunningham & Brookbank, 1988) shows the great range of life-spans to be found among contemporary species. From the selected values seen in Table 8.2 it should be apparent that the longest-lived species are to be found among plants, with such species as the Bristlecone pine, the Giant sequoia and certain desert creosotes living thousands of years. The longest-lived fish is a species of rockfish (*Sebastes aleutianus*) with a maximum recorded age of 140 years (Chilton & Beamish, 1982, cited by Beverton, 1987). In a shorter-lived species, the walleye of North America, there appears to be a positive correlation between increased latitude and longevity. Reproduction starts later in colder climates but fish live longer there and, in addition, are larger where ambient temperatures are lower. It is not known whether this is a general phenomenon among poikilotherms, but there seems to be little evidence for such temperature-related longevity

216   W. A. Stini

Table 8.3. *Brain size, body weight and life-span in tree shrews and five primate species*

Animal	Brain size (cm^3)	Body weight (g)	Life-span (years) Observed	Life-span (years) Predicted
Tree shrew	5.3	275	4	7.7
Rhesus macaque	106.0	8 719	29	27
Orang-utan	420.0	69 000	50	41
Chimpanzee	410.0	49 000	45	43
Gorilla	550.0	140 000	40	42
Homo sapiens	1446.0	65 000	95–100	92

From Cutler (1978).

differences among homeotherms. Interestingly, Beverton points outs that whether longevity is long or short in walleye populations, the total lifetime of egg production is similar. From this, he concludes that reproductive strategy is target-seeking whether reproduction starts early or late. This conclusion had also been reached earlier by Comfort (1963).

Cutler (1978) presented evidence that life-span can be predicted on the basis of body and brain weight. Comparisons of tree shrews and five primate species are shown in Table 8.3. A number of biologists have argued that longevity in mammals is proportional to relative brain weight (Mallouk, 1975; Economos, 1980a,b; Hofman, 1983, 1984). Sacher (1959, 1975, 1978) cited abundant evidence that maximal life-span is better correlated with brain weight than with body weight and that there is also a negative correlation between life-span and metabolic rate. A number of researchers have pointed out a generally positive relationship between body size and maximal life-span even when brain weight is not taken into consideration (Bonner, 1965; Sheldon, Prakash & Suttcliffe, 1972; Blueweiss et al., 1978). Ingram & Reynolds (1987) demonstrated that the correlation coefficient between log body weight and log life-span is about 0.79 for all mammals and about 0.86 for rodents, although, among humans, higher body weight is generally associated with decreased life-span. Lew & Garfinkel (1979) and Stunkard (1983) were the source of the human comparisons cited by Ingram & Reynolds. (Some additional comments on human body weight/life-span relationships will be found in a later section.)

Prothero & Jürgens (1987) used a power function to compare the life-span and body weight relationships of a number of mammals, extending the comparisons to include a number of organ weights in addition to brain weight. From these comparisons, it was found that kidney, liver and adrenal gland weights explain as much of the variance in life-span as does brain weight. When life-spans within a given weight range were compared, it was found that, while primates have a greater maximal life-span as related to body weight than all other non-flying mammals, their life-spans were shorter than bats of similar weight.

The relationships of size, metabolic rate and activity levels to life-span have long been of interest to researchers concerned with the determinants of longevity (Zepelin & Rechlschaffer, 1974; Boddington 1978; Lew & Garfinkel, 1979; Western, 1979; Western & Ssemakula, 1982; Stunkard, 1983; Latin, Johnson & Ruhling, 1987). Although the reasons for the reported disparities of life-span/body weight and life-span/organ weight relationships among mammalian species are not fully understood, they are consistent with intraspecific differences that have also been well documented. The life-spans of smaller varieties of domestic dogs exceed that of their larger conspecifics by nearly a factor of two, a difference that could also be predicted by the higher ratio of brain to body weight in the smaller varieties. Also in agreement with such predictions is the observed superiority of female life expectancies in mammalian species including *Homo sapiens* (Rockstein *et al.*, 1977). In addition to the greater adaptability conferred by a proportionally larger brain in a challenging environment (Moment, 1982), there are probably other factors associated with these intraspecific differences in maximal life-span. For example, the log body weight to log metabolic rate values for males and females reveal a relatively higher metabolic rate in males than in females. The inverse correlation between metabolic rate and life-span that has been generally observed could well be a factor in this difference. When the relationship between brain weight, body weight and life-span is viewed from the perspective of its part in the *K*-adapted strategy discussed earlier, a proportionally larger brain associated with an extended period of learning in a challenging environment can be viewed as an important component of the adaptive complex.

Genetics and aging

Although species differences in life-span can be understood, at least in part, as a result of the evolution of an adaptive complex, within-species variation in life-span provides considerable evidence of interaction between the genetic composition of individuals and the environment (Wright & Butler, 1987). In addition to the generally superior life expectancies of females, whatever the species, there are clear differences in the life expectancies of captive as compared to wild animals (Snyder & Moore, 1968; Jones, 1982). It is generally true that predation and accidents result in steadily decreasing survival probabilities in wild species where death attributable to senility is exceedingly rare. As a result, the maximum life-span is seen only in captive populations. Domestic animals bred for specific economically beneficial traits are a special category that may not accurately reflect the maximum life-span of the wild species from which they were derived.

Among humans, there is good evidence of genetic determinants of life-span. An early study comparing ages at death of monozygotic twins compared to those of dizygotic twins (Kallman & Sander, 1949) recorded an average intrapair difference in age at death of around 37 months for monozygotic pairs as opposed to 78 months for dizygotic pairs. More recent work by Bank & Jarvik (1978) confirms these results. The familial component of longevity has also been confirmed through the determination of the tendency of nonagenarians to descend from long-lived parents and grandparents (Abbot *et al.*, 1974; Murphy, 1978). However, the mode of inheritance is undoubtedly not a simple one and a large number of traits is potentially implicated in success or failure to survive to advanced age (McClearn, 1987).

Precocious aging, as seen in the Hutchinson–Gifford progeria syndrome provides an example of an inherited acceleration of senility. What makes this condition so striking is the pervasiveness of the aging changes that occur. During the first 11 years of life, patients develop generalized atheroslerosis with occlusion of the coronary arteries. They are therefore subject to strokes and heart attacks, with a majority of progeria patients dying of heart attacks at a median age of 12 years (Schneider & Reed, 1985). In addition, osteoporosis resembling that expected in the sixth or seventh decade of life can be found in these 12-year-olds. However, some features of aging, such as diabetes, cataracts, degenerative joint disease and Alzheimer's disease, and cancer are generally absent. Although rare – the occurrence of Hutchinson–Gifford progeria syndrome is of the order of one in every eight million births – these patients reveal much about the degree to which certain genetic determinants can control major aspects of the aging process.

Contemporary demographics: aging and mortality

Genetic predisposition to heart attacks and cancer does appear to play a role in limiting the life-span of a sizable proportion of contemporary human populations. However, interpopulational differences in mortality

Table 8.4. *Causes of mortality in Central European and Central American countries compared*

Country	Deaths per 100 000 of the population				
	Cardio-vascular disease	Cancer	Diabetes	Influenza and pneumonia	Infectious diseases of the gut
Hungary	714.1	299.0	17.9	14.9	0.2
Austria	658.4	279.2	16.1	26.6	0.1
Czechoslovakia	611.5	254.9	17.5	43.4	0.3
Mexico	98.0	33.7	21.7	47.8	50.7
Guatemala	52.3	30.7	5.2	144.0	203.9
El Salvador	59.2	17.1	5.4	13.6	46.5

Source: *United Nations Demographic Yearbook: 1984*. New York, 1986.

rates due to these causes have alerted researchers to the existence of environmental factors that can moderate or amplify inherited tendencies. Table 8.4 compares recent mortality rates in several European and Central American populations. The sharp difference in mortality attributable to cardiovascular disease and cancer in European as opposed to Central American populations reflects the substantial environmental component in the determination of life expectancies. It should be kept in mind, however, that lower mortalities due to cardiovascular disease and cancer in the Central American countries reflect a different demographic profile. Where a larger proportion of the population survives into and past midlife, the proportion of the population at risk for these age-related conditions increases commensurately (Stini, Harrington & Sumner, 1985; Stini, 1987). There is clear evidence of mortality trends in the United States within the twentieth century. Table 8.5 shows the recent decline in mortalities due to cardiovascular diseases since 1960. It appears that the mortality rate for cardiovascular diseases peaked sometime around 1955 in the United States, while cancer mortalities trend inexorably upward as the proportion of elderly citizens rises. The demographic profile of the United States in 1985 (sexes combined) shows the characteristically lengthened survival past the age of 50 associated with modern industrial societies (Fig. 8.1). Comparison with the demographic profile for Tunisia in 1984 (Fig. 8.2) illustrates the magnitude of differences in the proportion of older people in developed and developing countries. Whereas 26 % of the population of the United States is beyond the age of 50, the Tunisian percentage is roughly half of that at 13 %. Although the

Table 8.5. *Causes of mortality in the United States*

Cause of death	Deaths per 100 000 in year				
	1900	1960	1970	1980	1984–85
Cardiovascular disease	345.2	515.1	496.0	436.4	410.5
Cancer	64.0	149.2	162.8	183.9	193.2
Pneumonia and influenza	202.0	37.3	37.0	24.1	21.1
Tuberculosis	194.0	6.1	2.6	0.9	0.9
Diphtheria	40.0	NA	NA	NA	NA
Typhoid and paratyphoid	31.0	NA	NA	NA	NA
Measles	13.0	NA	NA	NA	NA

Source: US Bureau of the Census Statistical Abstract of the United States (106th edn): *1986*.
Washington, DC.
NA: Frequency too low to calculate.

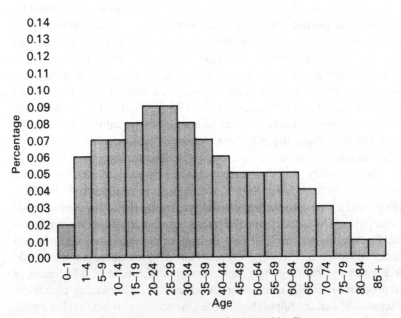

Fig. 8.1. USA 1985: age distributions (sexes combined).

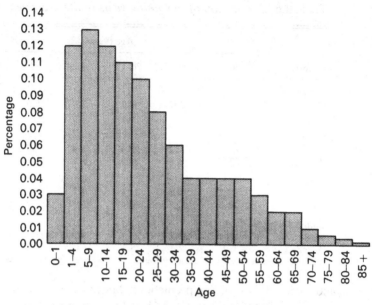

Fig. 8.2. Tunisia 1984: age distributions (sexes combined).

demographic profile of Tunisia in 1984 was not identical to that of the United States in 1900, the two are quite similar in being marked by high infant mortalities and high incidence of death due to infectious diseases. As a result of these factors, average life expectancy in Tunisia in 1984 was quite similar to that of the United States in 1900.

One of the most striking aspects of the contemporary demographic structure of the United States has been the steady increase of survivors in the most advanced age categories – often referred to as the 'old old' – those over 85 years of age. In 1989, this was the fastest-growing segment of the United States population and promises to remain so. As a result of this survival trend, the number of centenarians has steadily increased to the current level of about 40 000 individuals. It can be anticipated that the proportion of very old individuals will also increase in developing countries as infectious disease and other environmental causes of mortality recede in importance. Reference to Table 8.6 will show that, even in 1985, the expectation of remaining life in the more advanced age groups in developing countries resembled those in the United States and Japan to a surprising extent.

When the trend toward increased surviviorship in the developing countries is taken into account, forecasts of future population trends in

Table 8.6. *Expectation of remaining years of life at a given age*

Country	Age (years)					
	50	60	70	75	80	85
India						
Men	19.2	13.6	9.3	7.7	6.0	3.8
Women	19.7	13.8	9.5	7.8	6.0	3.8
Guatemala						
Men	23.0	16.2	10.4	8.4	6.0	4.3
Women	23.8	16.6	10.6	8.2	6.2	4.4
Japan						
Men	27.2	19.0	11.7	8.7	6.4	4.6
Women	31.7	22.7	14.4	10.8	7.7	5.3
United States						
Men	25.4	17.8	11.5	9.0	6.0	5.1
Women	31.0	22.6	15.2	11.0	9.0	6.6

Asia yield some rather sobering predictions. Martin (1988) predicts that the population of individuals over age 65 will double in India and China between 1980 and 2000 and will double again between 2000 and 2025. If this does occur, 12.9 % of the population of China in 2025 will be over the age of 65. In Japan, the proportion will be in excess of 20 %. Clearly, these increases in the elderly population with their associated demands on health care delivery systems throughout the world will create unprecedented social and economic pressures.

The level of survivorship through early and middle life seen in the contemporary Japanese population provides us with a model for the immediate future. Approached from the perspective of age-specific conditional probabilities of survival (see Fig. 8.3), it can be seen that the probability of death in any given year is less than 1 % from the age of 5 years until about the age of 45 for a male Japanese. Even beyond the age of 85, the probability of survival remains above 50 %. These high survival probabilities have led Williams & Taylor (1987) to point out that, in contemporary human populations, the first 10 minutes of life probably yield a higher mortality rate than that experienced by the average centenarian.

The role of body composition in survival

The importance of cardiovascular disease and cancer as causes of mortality has led to suspicion that obesity accelerates aging and raises the

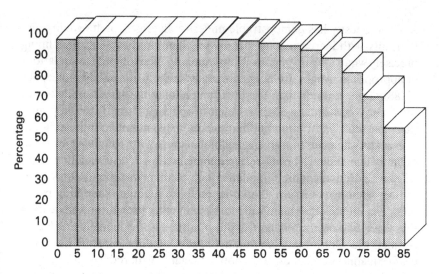

Fig. 8.3. Age-specific conditional probabilities of survival, Japanese males, 1984.

probability of premature death. There is considerable evidence to support the argument that dietary patterns associated with obesity increase the likelihood of atherosclerosis, diabetes and hypertension, and their concomitants of heart disease and stroke (Van Itallie, 1979; Hirsch, Bell & Dwyer, 1985). There is also abundant evidence that diets high in animal fats and low in fibre increase the risk of several types of cancer. The evidence that being overweight in comparison to recommended values such as the Metropolitan Life Insurance Recommended Weights for Height is less conclusive, however. The distribution as well as amount of body fat appears to have a role to play in predisposing the individual to heart disease and, perhaps, cancer as well. The tendency to accumulate omental fat is associated with cardiovascular disease to a far greater extent than the tendency to deposit fat in the limbs or subcutaneously. This is especially true in adult males. The general tendency for females to deposit fat in the limb and subcutaneous compartments is associated, whether directly or indirectly, with lower incidence of cardiovascular disease. There is considerable evidence, drawn from experimental work with rats, that increased fat cell size associated with obesity reduces the response to insulin (Lawrence *et al.*, 1989). Loss of ability to control blood sugar levels along with other consequences of insulin insensitivity has the potential of aggravating or even initiating tendencies toward cardiovascular disease.

Thus, obesity has demonstrable undesirable effects on morbidity and mortality with increasing age. However, there are problems in defining precisely what 'obesity' means. Consequently, blanket recommendations of appropriate weight for height are inadvisable. Ross and his colleagues (1988) argue strongly that weight recommendations based on the use of the Body Mass Index (BMI = weight (kg)/height^2 (m^2)) can be potentially harmful. They point out that exceptionally muscular individuals are consistently heavier for a given height than those with a higher proportion of fat to lean tissue. Therefore, a recommendation to lose weight through caloric restriction could, if implemented, result in loss of muscle tissue at a time when every effort should be made to maintain lean body mass. Ross's cautionary comments are consistent with observations made over the years by Andres (1980a,b 1985). On the basis of evidence collected through the Baltimore Longitudinal Study, Andres concluded that excessive leanness and obesity are both associated with reduced life expectancy, but that a weight greater than the recommended ideal weight for height is associated with greater than average longevity. Thus, the curve of mortality as related to BMI is U-shaped, with lowest mortalities occurring at BMI values which would fall into the category of moderately obese. Waaler has reported similar findings in a Norwegian population (Waaler 1984). A number of other investigators have found a positive relationship between greater body weights and life expectancy (Libow, 1974; Keys, 1980; Avons, Ducimetiere & Rakotovao, 1983).

In a retrospective study of over 8000 patients admitted to hospitals in Nebraska, Potter, Schafer & Bohi (1988) found that the U-shaped curve of mortality seen by Andres and Waaler could be successfully applied to a population of individuals hospitalized for serious illnesses including heart disease, cancer and pneumonia. In this study, it was seen that obesity was associated with higher mortality only when the patient was 100 % or more overweight. On the other hand, patients who were at or below ideal weight as estimated by calculation of the BMI were also subject to a higher than average risk of mortality. The deleterious effects of being extremely fat or extremely lean increased with age in this population (see Fig. 8.4). The curves of predicted mortality with relation to BMI (Fig. 8.5) were very similar whether admission was for cancer, cardiovascular disease or for pneumonia and other conditions.

The findings of Andres, Waaler, and Potter et al. raise a number of questions concerning the relationship of body weight to disease resistance and the ability to recover from serious illness. Superior chances of survival with moderate overweight are most pronounced in the oldest patients. This leads to the suspicion that nutrient reserves play an

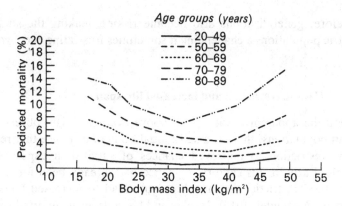

Fig. 8.4. Probability of death as a function of body mass index for each of five age groups. (From Potter *et al.*, 1988.)

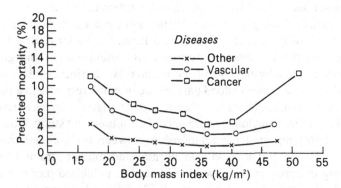

Fig. 8.5. Predicted probability of death as a function of body mass index for three disease categories. (From Potter *et al.*, 1988.)

increasingly important role in recuperative capacity as age increases. Although there are differences of opinion concerning some of the markers indicating declining reserves with age (Campion, DeLabry & Glynn, 1988), this would seem to be a promising area for future research on the predictors of long life. One thing is certain, however: human variability makes virtually any attempt at generalization risky. Moreover, individual differences become ever more pronounced as age increases.

Therefore, generalizations also become riskier, making the study of geriatric populations a challenging if sometimes frustrating endeavour.

Caloric restriction and increased life-span

Despite the apparent advantage of above-average weight observed in human populations, there is a substantial body of evidence that relates caloric restriction to improved chances of survival in experimental animals. Early in this century, evidence had already been found that severe food restriction in early life could lead to increased longevity (Osborne & Mendel, 1915). McCay and his colleagues reported similar findings 20 years later (McCay, 1935; McCay et al., 1939, 1941, 1952). A number of laboratories have extended this line of research to include work with several varieties of rats, mice, guinea pigs, and recently primates (Short, Williams & Bowden, 1987). Ross (1966) reported increased life expectancy by dietary modification in modern rats. His results went beyond those of Osborne and McCay, whose conclusions were essentially drawn from a perceived extension of life associated with prolongation of the growth and development process during early life. Later work of Ross & Bras (1965) and Ross, Lustbader & Bras (1976) focused on the incidence of tumours and alterations of the growth response effected by dietary restriction and related to increased life-span. Weindruch et al. (1986) added data on the immune response in a similar study. Goodrick (1974) called attention to the relationship between diet, exercise patterns and longevity in mice. In a subsequent study, Goodrick (1977, 1978) reported the relationship between body-weight changes over a lifetime and longevity in normal and mutant strains of mice differing in maximal body weight. In studies published more recently (Goodrick, 1980; Goodrick et al., 1982, 1983a,b), Goodrick and his collaborators reported the relationships between intermittent feeding, voluntary exercise, body weight and longevity in rats. In these studies, the positive association between the amount of exercise voluntarily undertaken and longevity raised the possibility that exercise was the intervening variable that linked dietary restriction to increased life-span in all of the earlier studies.

Other researchers (Leto, Kokkonen & Barrows, 1976; Barrows & Kokkonen, 1982) have explored the relationship of protein intake to life-span, concluding that a restricted diet made up of 4 % protein (as opposed to the standard 26 %) effectively reduced body weight in female mice and increased life-span. In this study, it was seen that rectal temperatures were higher throughout life in the protein-restricted animals. The question of the relationship of food intake and its effect on metabolic rate is of considerable interest in relation to life-span extension. The 'rate of living' hypothesis, linking metabolic levels to length of life, was discussed by Pearl over 60 years ago (Pearl, 1928). The relationship between metabolic rate and the presence of free radicals implicated in early cell death is an area of much ongoing research (Tolmasoff, Ono & Cutler, 1980) as is the role of antioxidants such as vitamin E in neutralizing free radicals (Housset, 1987).

Masoro, Yu & Bertrand (1982) also explored the question of the mechanism by which food restriction increases life-span. Their conclusion was that, when caloric intake and oxygen consumption are related to weight over a lifetime, no difference in life-span is seen between dietary restricted and *ad libitum* fed animals. However, this does not necessarily refute the rate of living hypothesis when the total caloric intake and oxygen consumption per animal (ignoring size) is related to life-span. The smaller animals live longer, consume fewer calories and less oxygen. Masoro and his collaborators continue to explore the relationships between aging and the metabolic rate as related to variations in food intake in rats (Masoro, 1984; McCarter, Masoro & Yu, 1985; Yu, Masoro & McMahan, 1985).

Despite the plethora of studies yielding evidence that dietary restriction of one kind or another can increase life-span in a number of mammalian species, demonstration of such life-span increase in higher primates is still lacking. Dietary practices that reduce the risk of cardiovascular disease, cancer and diabetes certainly have a beneficial effect in increasing life expectancy. There are strong arguments in favour of a low-fat, high-fibre diet. However, animal experiments linking extension of the life-span to reduced intake of calories and/or proteins do not at this time form a suitable basis for recommendations affecting human dietary practices. The U-shaped curve relating mortality to body weight in human populations should serve as a warning that there are many differences between a free-living population exhibiting high levels of variability, mobility and disease exposure and laboratory populations of inbred animals in a controlled environment.

Conclusions

Senescence and death are intrinsic aspects of life in higher organisms. They may be viewed as part of the price each species pays in order to retain the capacity to evolve. In species whose adaptive strategies have favoured survival of a high proportion of offspring into adulthood, there is also a tendency to exhibit a number of related attributes. Among these are a density-dependent mortality pattern, population size somewhere near the carrying capacity of the resource base, delayed reproduction, slow development, larger body size, longer life-span and competitive efficiency. These attributes can all be identified in the adaptive strategies of human populations everywhere. The degree to which human populations have been able to gain control over the natural environment and expand their resource base has led to relaxation of many of the constraints on population size and individual survival that prevailed over most of human history. The result has been a reduction in the degree to which extrinsic factors determine human life expectancy. As a result of this reduction, it is now of considerable interest to identify the intrinsic limitations on the length of human life.

World demographic trends are producing extensions of average life expectancy in developing countries similar to those that have occurred in the industrialized ones. In developing countries, infant mortalities are also being reduced, rates of child growth and maturation accelerated and body size increased. Attainment of the genetic potential for body size and reduction of the risk of premature death due to starvation, trauma and infectious disease is becoming a world-wide pattern. While humans are not the longest-lived species, evolution has produced a combination of body size, brain size and longevity that is unique. A large percentage of humans is presently achieving both its full potential for body size and length of life. Is there a possibility that intrinsic constraints on life-span can be relaxed? The best evidence so far would support the argument that the species' life-span is a trait of fundamental evolutionary significance. If this is so, anything beyond minor extensions of maximal life-span would be unlikely, although premature deaths may be reduced to a minimum. Systems and processes of fundamental biological importance are characterized by redundant back-up systems. Although the effort to extend the human life-span to the maximum may ultimately be destined for frustration, the discovery of the nature of the redundant systems that guarantee human death should be of considerable practical and theoretical significance.

In summary, human populations are experiencing substantial increases in average life expectancy. However, it does not appear that the increase in average life expectancy is an indication of an increase in maximum life expectancy. Senescence and death are believed to have evolved after the emergence of eukaryotes and have been retained in all subsequent phyletic lines. There is, however, considerable variation in the maximum life-span characterizing various species. The human life-span is among the longest among homeotherms. Maximum life-span in humans is in general agreement with predictions made on the basis of the ratio of brain size to body size in mammals. The process of aging and senescence involves a number of physiological changes which have been identified but which are not fully understood. Decline of the immune system, accumulation of by-products of metabolism, failure of DNA repair mechanisms, and a slowing of metabolic rate all play a role in the aging process. There is substantial evidence that there are genetic determinants of longevity as seen in comparisons of monozygotic and dizygotic twins. It is generally true that individuals who attain great age have at least one parent who also did. Predispositions to the major cause of mortality, heart disease, and cancer can be mitigated in a variety of ways. Studies of the relationship of body composition to life expectancy have raised as many questions as they have answered. Increases in life-span of experimental animals through dietary restriction have raised questions concerning the possibility of reducing human morbidity and mortality through reduced consumption of protein and/or calories. However, comparison of rates of survival in humans hospitalized for a variety of causes indicates that the highest risk of mortality is experienced by the individual who is under rather than over ideal weight for height.

References

Abbott, M. H., Murphy, E. A., Bolling, D. R. & Abbey, H. (1974). The 'familial component' in longevity. A study of offspring of nonagenarians. *Johns Hopkins Medical Journal*, **134**, 1–16.

Andres, R. (1980a). Effect of obesity on total mortality. *International Journal of Obesity*, **4**, 381–6.

(1980b). Influence of obesity on longevity in the aged. In *Aging, Cancer and Cell Membranes*, ed. C. Borek, C. M. Fenoglio & D. W. King, pp. 238–46. New York: George Thieme Verlag.

(1985). Mortality and obesity: the rationale for age-specific height–weight tables. In *Principles of Geriatric Medicine*, ed. B. Andres, E. L. Bierman & W. R. Hazzard, pp. 311–18. New York: McGraw-Hill.

Arking, R. (1987). Genetic and environmental determinants of longevity in *Drosophila*. In *Evolution of Longevity in Animals*, ed. A. E. Woodhead & K. H. Thompson, pp. 1–22. New York: Plenum Press.

Avons, P., Ducimetiere, P. & Rakotovao, R. (1983). Weight and mortality. *Lancet*, **1**, 1104.

Balin, A. K. (1982). Testing the free radical theory of aging. In *Testing the Theories of Aging*, ed. R. C. Adelman & G. S. Roth, pp. 137–83. Boca Raton: CRC Press.

230   W. A. Stini

Bank, L. & Jarvik, L. F. (1978). A longitudinal study of aging in human twins. In *Genetics of Aging*, ed. E. L. Schneider, pp. 303–33. New York: Plenum Press.

Barrows, C. H. & Kokkonen, G. C. (1982). Dietary restriction and life extension – biological mechanisms. In *Nutritional Approaches to Aging Research*, ed. G. B. Moment, pp. 219–43. Florida: CRC Press.

Bergmann, K., Walford, R. L., Esra, G. N. & Hall, K. Y. (1981). A correlation of repair and UV-induced DNA damage and maximum lifespan in primate lymphocytes. In *Thematic Sessions XII International Congress of Gerontology*, 2, 125–38.

Bernstein, H. (1977). Germ line recombination may be primarily a manifestation of DNA repair processes. *Journal of Theoretical Biology*, 69, 371–80.

Beverton, R. V. H. (1987). Longevity in fish: some ecological and evolutionary considerations. In *Evolution of Longevity in Animals*, ed. A. D. Woodhead & K. H. Thompson, pp. 161–85. New York: Plenum Press.

Blueweiss, L., Fox, H., Kudzma, V., Nakaahima, D., Peters, R. & Sams, S. (1978). Relationships between body size and some life history parameters. *Oecologia* (Berlin), 37, 257–72.

Boddington, M. J. (1978). An absolute metabolic scope for activity. *Journal of Theoretical Biology*, 75, 443–9.

Bonner, J. T. (1965). *Size and Cycle*. Princeton, NJ: Princeton University Press.

Burmer, G. C., Motulsky, H., Ziegler, C. J. & Norwood, T. H. (1983). Inhibition of DNA synthesis in young cycling human diploid fibroblast-like cells upon fusion to enucleate cytoplasts from senescent cells. *Experiments in Cell Research*, 145, 79–84.

Campion, E. W., DeLabry, L. O. & Glynn, R. J. (1988). The effect of age on serum albumin in healthy males: report from the normative aging study. *Journal of Gerontology: Medical Sciences*, 43/1, M18–20.

Chilton, D. E. & Beamish, R. J. (1982). Age determination methods for fishes studied by the groundfish program at the Pacific Biological Station. *Canadian Special Publications in Fisheries and Aquatic Science*, 60, Ottawa: Ministry of Fisheries & Aquatic Resources.

Comfort, A. (1963). Effect of delayed and resumed growth on the longevity of a fish (*Lebistes reticulatus*, Peters) in captivity. *Gerontologia*, 8, 150–9.

Cunningham, W. R. & Brookbank, J. W. (1988). *Gerontology: the psychology, biology and sociology of aging*, pp. 28–9. New York: Harper & Row.

Cutler, R. G. (1975). Evolution of human longevity and the genetic complexity governing aging rate. *Proceedings of the National Academy of Sciences, USA*, 72, 4664–8.

(1978). The evolutionary biology of senescence. In *The Biology of Aging*, ed. J. A. Behnke, C. E. Finch & G. B. Moment, pp. 311–60. New York: Plenum Press.

(1984). Evolutionary biology of aging and longevity in mammalian species. In *Aging and Cell Function*, ed. J. E. Johnson Jr, pp. 1–147. New York: Plenum Press.

Drescher-Lincoln, C. K. & Smith, J. R. (1983). Inhibition of DNA synthesis in senescent proliferating human diploid fibroblasts by fusion with senescent cytoplasts. *Experiments in Cell Research*, 144, 455–62.

(1984). Inhibition of DNA synthesis in senescent proliferating human cybrids is mediated by endogenous proteins. *Experiments in Cell Research*, **153**, 208–17.

Economos, A. C. (1980a). Taxonomic differences in the mammalian lifespan–body weight relationship and the problem of brain weight. *Gerontology*, **26**, 90–8.

(1980b). Brain–lifespan conjecture: a re-evaluation of the evidence. *Gerontology*, **26**, 82–9.

Francis, A. A., Lee, W. H. & Regan, J. D. (1981). The relationships of DNA excision-repair of UV-induced lesions to the maximum lifespan of mammals. *Mechanisms of Aging and Development*, **16**, 181–94.

Goodrick, C. L. (1974). Effects of exercise on longevity and behavior of hybrid mice which differ in coat color. *Journal of Gerontology*, **29**, 129–33.

(1977). Body weight change over lifespan and longevity for *C57BL/6J* mice and mutations which differ in maximal body weight. *Gerontology*, **23**, 405–13.

(1978). Body weight increment and length of life: the effect of genetic constitution and dietary protein. *Journal of Gerontology*, **33**, 184–90.

(1980). Effects of long-term voluntary wheel exercise on male and female Wistar rats, 1. Longevity, body weight, and metabolic rate. *Gerontology*, **26**, 22–33.

Goodrick, C. L., Ingram, D. K., Reynolds, M. A., Freeman, J. R. & Cider, N. L. (1982). Effects of intermittent feeding upon growth and lifespan in rats. *Gerontology*, **26**, 233–41.

(1983a). Differential effects of intermittent feeding and voluntary exercise on body weight and survival in adult rats. *Journal of Gerontology*, **38**, 36–45.

(1983b). Effects of intermittent feeding upon growth activity and lifespan in rats allowed voluntary exercise. *Experiments in Aging Research*, **9**, 203–9.

Grigliatti, T. A. (1987). Programmed cell death and aging in *Drosophila melanogaster*. In *Evolution of Longevity in Animals*, ed. A. D. Woodhead & K. H. Thompson, pp. 193–206. New York: Plenum Press.

Haldane, J. B. S. (1941). *New Paths of Genetics*. London: Allen & Unwin.

Hamburger, V. (1958). Regression versus peripheral control of differentiation in motor hypoplasia. *American Journal of Anatomy*, **102**, 365–409.

Harman, D. (1981). The aging process. *Proceedings of the National Academy of Science, USA*, **78**, 7124–8.

Hart, R. W. & Turturro, A. (1983). Theories of aging. In *Review of Biological Research in Aging, Vol. 1*, ed. M. Rothstein, pp. 5–17. New York: Alan R. Liss.

Hayflick, L. (1985). Theories of biological aging. *Experimental Gerontology*, **29**, 145–59.

Hirsch, G. P. (1978). Somatic mutations and aging. In *The Genetics of Aging*, ed. E. L. Schneider, pp. 91–134. New York: Plenum Press.

Hirsch, H. R. (1987). Why should senescence evolve? An answer based on a simple demographic model. In *Evolution of Longevity in Animals*, ed. A. D. Woodhead & K. H. Thompson, pp. 75–90. New York: Plenum Press.

Hirsch, J., Bell, C. H. & Dwyer, J. T. (1985). Health implications of obesity. National Institutes of Health consensus development conference statement. *Annals of Internal Medicine*, **103**, 147–51.

232   W. A. Stini

Hofman, M. A. (1983). Energy metabolism, brain size and longevity in mammals. *Quarterly Review of Biology*, **58**, 495–512.
(1984). On the presumed co-evolution of brain size and longevity in hominids. *Journal of Human Evolution*, **13**, 371–6.
Housset, B. (1987). Biochemical aspects of free radicals metabolism. *Clinical Respiratory Physiology*, **23**, 287–90.
Ide, T., Yoshiaki, T., Ishibashi, S. & Mitsui, Y. (1983). Re-initiation of host DNA synthesis in senescent human diploid cells by infection with simian virus 40. *Experimental Cell Research*, **143**, 343–9.
Ingram, D. K. (1983). Toward the behavioral assessment of biological aging in the laboratory mouse: concepts, terminology, and objectives. *Experimental Aging Research*, **9**, 225–38.
Ingram, D. K. & Reynolds, M. A. (1987). The relationship of body weight to longevity within laboratory rodent species. In *Evolution of Longevity in Animals*, ed. A. D. Woodhead & K. H. Thompson, pp. 247–82. New York: Plenum Press.
Johnson, T. E. (1988). Mini-review: Genetic specification of life span: processes, problems, and potentials. *Journal of Gerontology: Biological Sciences*, **43/4**, B87–92.
Jones, M. L. (1982). Longevity of captive animals. *Zoological Garden, NF Jena*, **52**, 113–28.
Kallman, F. J. & Sander, G. (1949). Twin studies on senescence. *American Journal of Psychiatry*, **106**, 29–36.
Keys, A. (1980). Overweight, obesity, coronary heart disease, and mortality. *Nutrition Reviews*, **38**, 297–307.
Kirkwood, T. B. L. (1977). Evolution of aging. *Nature*, **270**, 301–4.
(1981). Repair and its evolution: survival versus reproduction. In *Physiological Ecology: an evolutionary approach to resource use*, ed. C. R. Townsend & P. Calow, pp. 165–89. Oxford: Blackwell Scientific Publications.
(1985). Comparative and evolutionary aspects of longevity. In *Handbook of the Biology of Aging (2nd edn)*, ed. C. E. Finch & E. L. Schneider, pp. 27–44. New York: van Nostrand Reinhold.
(1987). Immortality of the germ-line versus disposability of the soma. In *Evolution of Longevity in Animals*, ed. A. D. Woodhead & K. H. Thompson, pp. 209–18. New York, Plenum Press.
(1989). DNA mutations and aging. *Mutation Research*, **219**, 1–7.
Kirkwood, T. B. L. & Cremer, T. (1982). Cytogerontology since 1881: a reappraisal of August Weismann and a review of modern progress. *Human Genetics*, **60**, 101–21.
Latin, R. W., Johnson, S. C. & Ruhling, R. O. (1987). An anthropometric estimation of body composition of older men. *Journal of Gerontology*, **42/1**, 24–8.
Lawrence, J. C. Jr, Calvin, J., Cartee, G. D. & Holloszy, J. O. (1989). Effects of aging and exercise on insulin action in rat adipocytes are correlated with changes in fat cell volume. *Journal of Gerontology: Biological Sciences*, **44/4**, B88–92.

Leto, S., Kokkonen, G. C. & Barrows, C. H. (1976). Dietary protein, lifespan, and biochemical variables in female mice. *Journal of Gerontology: Biological Sciences*, **31**, 144–8.

Lew, E. A. & Garfinkel, J. (1979). Variations in mortality by weight among 750 000 men and women. *Journal of Chronic Disease*, **32**, 563–76.

Libow, L. S. (1974). Interaction of medical, biologic, and behavioral factors on aging, adaptation, and survival. An 11-year longitudinal study. *Geriatrics*, **29**, 75–88.

McCarter, R., Masoro, E. J. & Yu, B. P. (1985). Does food restriction retard aging by reducing the metabolic rate? *American Journal of Physiology*, **248**, E488–90.

McCay, C. M. (1935). The effect of retarded growth upon the length of lifespan and upon the ultimate body size. *Journal of Nutrition*, **10**, 63–79.

McCay, C. M., Lovelace, E., Sperling, G., Barnes, L. L., Litt, C. H., Smith, C. A. H. & Saxton, J. A. Jr (1952). Age changes in relation to the ingestion of milk, water, coffee, and sugar solutions. *Journal of Gerontology*, **7**, 161–72.

McCay, C. M., Maynard, L. A., Sperling, G. & Osgood, H. (1939). Retarded growth, lifespan, ultimate body size, and age changes in the albino rat after feeding diets restricted in calories. *Journal of Nutrition*, **18**, 1–13.

(1941). Nutritional requirements during the latter half of life. *Journal of Nutrition*, **21**, 45–60.

McClearn, G. E. (1987). The many genetics of aging. In *Evolution of Longevity in Animals*, ed. A. D. Woodhead & K. H. Thompson, pp. 135–44. New York: Plenum Press.

Mallouk, R. S. (1975). Longevity in vertebrates is proportional to relative brain weight. *Federation Proceedings*, **34**, 2102–3.

Martin, L. G. (1988). The aging of Asia. *Journal of Gerontology: Social Sciences*, **434**, S99–113.

Masoro, E. J. (1984). Food restriction and the aging process. *Journal of the American Geriatric Society*, **32**, 296–300.

Masoro, E. J., Yu, B. P. & Bertrand, H. A. (1982). Action of food restriction in delaying the aging process. *Proceedings of the National Academy of Science, USA*, **79**, 4239–41.

Medawar, P. B. (1952). *An Unsolved Problem in Biology*. London: H. K. Lewis.

Moment, G. B. (1982). Theories of aging: an overview. In *Testing the Theories of Aging*, ed, R. C. Adelman & G. S. Roth, pp. 1–23. Boca Raton: CRC Press.

Murphy, E. A. (1978) Genetics of longevity in man. In *The Genetics of Aging*, ed. E. L. Schneider, pp. 261–301. New York: Plenum Press.

Nanney, D. L. (1974). Aging and long-term temporal regulation in ciliated protozoa: A critical review. *Mechanisms of Aging and Development*, **3**, 81–105.

Norwood, T. H., Pendergrass, W. R., Sprague, C. A. & Martin, G. M. (1974). Dominance of the senescent phenotype in heterokaryons between replicative and post replicative human fibroblast-like cells. *Proceedings of the National Academy of Science, USA*, **71**, 2231–4.

Orgel, L. E. (1973). Aging clones of mammalian cells. *Nature*, **243**, 4415.

Osborne, T. B. & Mendel, L. B. (1915). The resumption of growth after continued failure to grow. *Journal of Biological Chemistry*, **23**, 439–47.

Pearl, R. (1928). *The Rate of Living.* London: University of London Press.
Periera-Smith, O. M., Fisher, S. F. & Smith, J. R. (1985). Senescent and quiescent cell-inhibitors of DNA synthesis: membrane-associated proteins. *Experiments in Cell Research*, 160, 297–306.
Pianka, E. R. (1974). *Evolutionary Ecology.* New York: Harper & Row.
Potter, J. F., Schafer, D. F. & Bohi, R. L. (1988). In-hospital mortality as a function of body mass index: an age-dependent variable. *Journal of Gerontology: Medical Sciences*, 43/3, M59–63.
Prothero, J. & Jürgens, K. D. (1987). Scaling of maximal lifespan in mammals. In *Evolution of Longevity in Animals*, ed. A. D. Woodhead & K. H. Thompson, pp. 49–74. New York: Plenum Press.
Rockstein, M., Chesky, J. A. & Sussman, M. L. (1977). Comparative biology and evolution of aging. In *Handbook of the Biology of Aging*, ed. C. E. Finch & L. Hayflick, pp. 3–34. New York: van Nostrand Reinhold.
Ross, M. H. (1966). Life expectancy modification by change in dietary regimen of mature rat. *Proceedings of the 7th International Congress of Nutrition*, 5, 35–8.
Ross, M. H. & Bras, G. (1965). Tumor incidence patterns and nutrition in the rat. *Journal of Nutrition*, 87, 245–60.
Ross, M. H., Lustbader, E. & Bras, G. (1976). Dietary practices and growth responses as predictors of longevity. *Nature*, 262, 548–53.
Ross, W. D., Crawford, S. M., Kerr, D. A., Ward, R., Bailey, D. A. & Mirwald, R. M. (1988). Relationship of the Body Mass Index with skinfolds, girths, and bone breadths in Canadian men and women, aged 20–70 years. *American Journal of Physical Anthropology*, 77/2, 169–73.
Russell, R. L. & Seppa, R. I. (1987). Genetic and environmental manipulation of aging in *Caenorhabditis elegans*, In *Evolution of Longevity in Animals*, ed. A. D. Woodhead & K. H. Thompson, pp. 35–48. New York, Plenum Press.
Sacher, G. A. (1959). Relation of lifespan to brain weight and body weight in mammals. In *Ciba Foundation Colloquia on Aging, Vol. 5, The Lifespan of Animals*, ed. G. A. W. Walstenholme & M. O'Connor, pp. 115–33. London: Churchill.
(1975). Maturation and longevity in relation to cranial capacity in hominid evolution. In *Primate Functional Morphology and Evolution*, ed. R. H. Tuttle, pp. 417–41. The Hague: Mouton.
(1978). Longevity and aging in vertebrate evolution. *Bioscience*, 28, 497–501.
Saunders, J. W. Jr (1956). Death in embryonic systems. *Science*, 154, 604–12.
Schneider, E. L. & Reed, J. D. (1985). Modulation of aging processes. In *Handbook of the Biology of Aging*, ed. C. E. Finch & E. L. Schneider, pp. 45–76. New York: van Nostrand Reinhold.
Sheldon, R. W., Prakash, A. & Suttcliffe, W. H. (1972). The size distribution of particles in the ocean. *Limnology and Oceanography*, 17, 327–40.
Sheldrake, A. R. (1974). The aging, growth and death of cells. *Nature*, 250, 381–5.
Short, R., Williams, D. D. & Bowden, D. M. (1987). Cross-sectional evaluation of potential biological markers of aging in pigtailed macaques: effects of age, sex, and diet. *Journal of Gerontology*, 42/6, 644–54.

Smith, J. R., Spiering, A. L. & Periera-Smith, O. M. (1987). Is cellular senescence genetically programmed? In *Evolution of Longevity in Animals*, ed. A. D. Woodhead & K. H. Thompson, pp. 283–94. New York: Plenum Press.

Smith-Sonneborn, J. (1987). Longevity in the Protozoa. In *Evolution of Longevity in Animals*, ed. A. D. Woodhead & K. H. Thompson, pp. 101–9. New York: Plenum Press.

Snyder, R. L. & Moore, S. C. (1968). Longevity of captive animals in Philadelphia Zoo. *International Zoo Yearbook*, **8**, 175–82.

Sohal, R. S. & Runnels, J. H. (1986). Effect of experimentally prolonged life span on flight performance of houseflies. *Experimental Gerontology*, **21**, 509–14.

Stein, G. H. & Yanishevsky, R. M. (1979). Entry into 'S' phase is exhibited in two immortal cell lines fused to senescent human diploid cells. *Experiments in Cell Research*, **129**, 155–65.

Stini, W. A. (1987). The demographics and the epidemiology of aging. *Collegium Antropologicum*, **11/1**, 3–14.

Stini, W. A., Harrington, R. J. & Sumner, D. R. (1985). Aging, changes in body composition, and the emergence of new morbidity patterns in industrialized nations. *Humanbiologie Budapest*, **16**, 179–95.

Strehler, B. L. (1977). *Time, Cells, and Aging* (2nd edn). New York: Academic Press.

Stunkard, A. J. (1983). Nutrition, aging, and obesity: a critical review of a complex relationship. *International Journal of Obesity*, **7**, 201–20.

Tolmasoff, J. M., Ono, T. & Cutler, R. G. (1980). Superoxide dismutose: correlation with lifespan and specific metabolic rate in primate species. *Proceedings of the National Academy of Science, USA*, **77**, 2777–81.

Van Itallie, T. B. (1979). Obesity: adverse effects on health and longevity. *American Journal of Clinical Nutrition*, **32**, 2723–33.

Waaler, H. T. (1984). Height, weight and mortality: the Norwegian experience. *Acta Medica Scandinavica, Supplement*, **679**, 1–56.

Weindruch, R., Walford, R., Flegiel, S. & Guthrie, D. (1986). The retardation of aging in mice by dietary restriction: longevity, cancer, immunity, and lifetime energy intake. *Journal of Nutrition*, **116**, 641–54.

Weismann, A. (ed.) (1889). *Essays upon Heredity and Kindred Biological Problems*, Oxford: Oxford University Press.

Western, D. (1979). Size, life history, and ecology in mammals. *African Journal of Ecology*, **17**, 185–204.

Western D. & Ssemakula, J. (1982). Life history patterns in birds and mammals and their evolutionary interpretation. *Oecologia (Berlin)*, **54**, 281–90.

Williams, G. C. (1957). Pleiotrophy, natural selection, and the evolution of senescence. *Evolution*, **11**, 398–411.

Williams, G. C. & Taylor, P. D. (1987). Demographic consequences of natural selection. In *Evolution of Longevity in Animals*, ed. A. D. Woodhead & K. H. Thompson, pp. 235–45. New York: Plenum Press.

Wright, B. E. & Butler, M. H. (1987). The heredity–environment continuum: a systems analysis. In *Evolution of Longevity in Animals*, ed. A. D. Woodhead & K. H. Thompson, pp. 111–22. New York: Plenum Press.

Yu, B. P., Masoro, E. J. & McMahan, C. A. (1985). Nutritional influence on aging of Fischer 344 rats: I. Physical, metabolic and longevity characteristics. *Journal of Gerontology*, **40**, 657–70.
Zepelin, H. Y. & Rechlschaffer, A. (1974). Mammalian sleep, longevity and energy metabolism. *Brain Behavior and Evolution*, **10**, 425–70

Index

Note: page numbers in *italics* refer to figures and tables